3/96

Lubrication for Industry

Kenneth E. Bannister

Lubrication for Industry

Industrial Press, Inc.

Library of Congress Cataloging in Publication Data

Bannister, Kenneth E., 1955–
    Lubrication for industry / Kenneth E. Bannister. — 1st ed.
    160 p. 15.6 × 23.5 cm.
    Includes bibliographical references and index.
    ISBN 0-8311-3061-X
    1. Lubrication and lubricants. I. Title.
  TJ1075.B284 1995
  621.8′9—dc20   95-6722
                                                                                CIP

Industrial Press Inc.
200 Madison Avenue
New York, New York 10016-4078

First Edition

Lubrication for Industry

First Printing

10   9   8   7   6   5   4   3   2   1

For Jo, David, and Jamie

Contents

Preface

In this book I have attempted to give the reader a fundamental understanding of lubrication as a whole. I have focused on the practical, daily aspects of lubrication as they pertain to industrial equipment maintenance. Lubrication is a vast subject—often very technical—with sometimes disastrous implications when misunderstood.

The book has been broken down into the following four main areas.

1. Chapters 1 and 2 explore *why* there is a need for lubrication.

2. Chapters 3 and 4 cover the *what* and *how much* of lubrication.

3. Chapter 5 addresses *how* to lubricate.

4. Chapters 6, 7, and 8 show *how* to take care of the lubrication systems and lubricants.

This book serves to overcome common misconceptions and acts as an overall reference guide for all who are in any way involved with lubrication.

Acknowledgements

Throughout the years, one looks back and discovers that without the help of certain people, life—both professional and personal—could have turned out very differently.

I am grateful for this opportunity to extend my warmest gratitude to my wife, children, and parents, who have always been there for me personally.

In my professional life, I would like to thank three people, in particular, who have been mentors to me: Norman Prescott, my high school teacher who first instilled the engineering spark within me: Ted Mankiewitz, a mentor who taught me the meaning of good engineering design; and Chris Smith, a good friend and excellent engineer who instilled confidence within me many years ago, and continues to be one of my "sounding boards."

Finally, I extend a special thanks to Jo Bannister for compiling the book. Thanks also to Troy Hoffman, to Engtech Industries Inc., and to all the companies noted who have given permission to reproduce their material within this book.

Lubrication for Industry

CHAPTER 1

Maintaining for Profit

1.1 Cost of Maintenance

Maintenance is BIG BUSINESS. In 1979, it was estimated that over 200 billion U.S. dollars were spent on maintenance in North America. Even more astounding, approximately one-third of this expenditure was determined to be unnecessary.

Since 1979, maintenance costs have escalated between 10% and 15% per annum. In 1990's terms, that places maintenance costs at over a half-trillion U.S. dollars—with unnecessary expenditures at close to $200 billion!

Maintenance is one of the few remaining areas of a company's expenditure that can be drastically improved upon. The Dupont Company recently stated that "maintenance is the single largest controllable cost within a plant." Maintenance costs are paid for directly out of the company profits and, when left unharnessed, can be truly "variable" in nature. In order to turn these volatile "variable" costs into "fixed" costs, we need to use structured controls and practices. This equates to the formation and utilization of a structured PM program.

PM is a common acronym for Preventive Maintenance, Predictive Maintenance, Proactive Maintenance, Planned Maintenance, or Productive Maintenance—whichever you choose, they all include lubrication as an important part of the maintenance procedure.

Recently, the use of internationalized initiatives to boost profits has given the term PM a new meaning.

New Meaning:

PM = Profit Maintenance.

It is generally accepted within the lubrication community that over 60% of all mechanical failures relate directly to poor or im-

Cost of Maintenance

 = Demand

VS

 = P.M.

Fig. 1.1. Cost comparison of demand versus PM-style maintenance.

proper lubrication practices. It is therefore easy to understand why a good PM program relies heavily on good lubrication practices. In 1989, an automotive industry study (see Fig. 1.1) concluded that in comparison to demand (fire-fighting) maintenance, PM was one-third of the cost.

Special Note: It is interesting to note that the lubrication industry has long claimed that the use of proper lubrication can effectively triple the life of a mechanical component.

1.2 Reasons for Equipment Failure

There are many reasons for equipment failure. These failures can be split into the following two specific categories:

1. failures that pertain to maintenance practices

2. failures that pertain to outside or other influences.

Fig. 1.2a points to the number one reason for equipment failure as being improper lubrication, when viewed from a maintenance standpoint.

The STLE (Society of Tribologists and Lubrication Engineers) recognizes and states that approximately 50% of bearing failures are due to *abrasion* or, simply put, "lack of lubrication."

MAINTENANCE

- Poor lubrication practices
- Faulty repairs
- Slow response time
- Lack of training
- Ineffective PM
- Inadequate Routine Maintenance

Fig. 1.2a. Reasons for equipment failure due to maintenance practices.

Fig. 1.2b shows the type of failures that occur which the maintenance department usually has little or no control over.

History has shown that with new equipment purchases, a good lubrication system has typically always been thought of as an expendable option. The importance of good lubrication was not usually realized until a failure occurred (if it was ever realized at all). There have been countless recorded instances where million-dollar capital investments have been lost prematurely because oil and grease nipples were thought to be adequate replacements for centralized automatic lubrication systems. A centralized lubrication system should be viewed as the "heart" that pumps lifeblood (oil or grease) into each component of the machine. Unfortunately, this is not always the attitude that holds true today.

Because of today's complex equipment and types of available lu-

OTHER

- Operator error
- Improper setup
- Poor material specification
- Sabotage
- Machine design
- Poor environment
- Poor training
- Improper application
- Poor housekeeping

Fig. 1.2b. Reasons for equipment failure due to outside influences.

bricants, the task of lubricating correctly calls for skill, initiative, and (above all) responsibility. This applies to equipment design engineers, purchasing, management, and maintenance staff.

· Equipment should be engineered to be "maintenance ready"—with centralized lubrication delivery systems "embedded" into the machine, complete with specific lubricant-type recommendations.

· Lubricant purchasing patterns should be consolidated and controlled—lubricants should be purchased on specification, *not* on price. (Too many lubricants are purchased without the benefit of input from the maintenance and engineering departments.)

· Management and maintenance staff have to adopt, and commit to, policies that will ensure proper lubrication techniques and training are utilized. This is especially important as more companies practice Total Productive Maintenance (TPM) methods, where the lubrication function and responsibility is passed along to the machine operator.

1.3 Benefits of Good Lubrication

In 1966, a landmark study was tabled to the British Government by H. Peter Jost (the Jost Report). This study was funded by the British Ministry of State for Education and Science. The committee, headed by Peter Jost, was asked to consider the position of lubrication education and research in the U.K., and to give an opinion on the needs of industry in this field. This was to be the world's first comprehensive study of how friction, lubrication, and wear directly and indirectly affected the country's Gross National Product in the areas of industry, natural resources, and agriculture.

A remarkable set of figures (see Fig. 1.3) relating directly to the savings accrued through improved lubricating practices were published. Because varying factors and conditions made accurate assessment impossible, a conservative rating scheme was used. It was generally recognized that these potential savings figures were undervalued. When equated to an individual company's spread-

· Reduction in energy consumption through lower friction	7.5%
· Savings in lubricant costs	20%
· Savings in maintenance repair and replacement costs	20%
· Savings in consequential losses due to downtime—depended on type of failure and type of industry (see note below)	Variable %
· Savings in investment due to higher utilization ratios and greater machine efficiency	1%
· Savings in investment through increased life	5% of new expenditure
· Savings in manpower	0.13%

Fig. 1.3. Jost Report estimates of the effect of improved tribology.

sheet, these percentages suddenly turn into a substantial potential profit picture.

Since 1966, there have been numerous other studies performed in the Western World that include the U.S., Canada, and Germany. All of these studies have mirrored the findings of the original "Jost Report."

Note: This type of savings is indirect and is dependent on the "cost of downtime" figure associated with each specific equipment piece or train. These figures will equate substantially higher than the repair itself.

Review Questions

1. What percentage of maintenance expenditure is deemed unnecessary?

2. How are maintenance costs paid for?

3. Name six meanings for PM.

4. What is the cost ratio of demand maintenance versus PM?

5. What are the two main categories associated with equipment failure?

6. Name three associated failures in each equipment failure category.

7. Based on the potential savings estimates for improved tribology, what are your company's potential savings?

CHAPTER 2

Lubrication Theory

2.1 Tribology

The term "tribology" owes its existence to the 1966 "Jost Report." In that report, the word tribology was used for the first time and was defined as:

> the science and technology of interacting surfaces in relative motion, and of the practices related thereto.

The term "tribology" relates directly to the combination of all sciences and technologies associated with the three areas of *friction, lubrication*, and *wear*, and is specifically concerned with the following elements:

- design and manufacture of any surface where two surfaces interact,

- the correct use of materials where solid materials interact,

- the interaction of wear surfaces with lubricants, and

- the reduction of friction and wear.

Tribology is a generic technology: it is a technology that yields national benefit. These benefits are great, but are diffused over a large number of recipients. Chapter 1 showed the substantial economic benefits that individual companies can attain through improved tribology.

In terms of criticality, tribology affects industry in the following four areas.

1. Tribology is critical to the reliability of mechanical and certain electrical products.

2. Tribology is critical to the efficiency of mechanical and certain electrical products.

3. Tribology directly affects the "maintainability" of a product.

4. Tribology affects a company's financial statement both directly and indirectly.

2.2 Defining Friction

In order to understand the basics of lubrication, it is fundamentally important to understand the reason why lubrication needs to exist at all.

Special Note: Lubrication must exist because of the force called friction.

Definition of **Friction** *(Webster's Dictionary Definition):* The force which opposes the movement of one surface sliding or rolling over another with which it is in contact.

Simply put, friction is the force that retards bodies in motion. When dealing with mechanical equipment, frictional force is all too obvious, but when attempting to deal with friction, we quickly realize it isn't such a simple force. It is known as the number one enemy to the machine. There are two unchanging, fundamental laws that apply to friction:

1. friction varies directly with load

2. friction is independent of surface area.

Special Note: It is generally accepted that over ⅓ of the world's energy production is consumed in overcoming friction!

The primary function of a lubricant is to help overcome friction. When two nonlubricated surfaces come into contact, an external force is required to move them over one another. Fig. 2.2a shows an elevation cross section of how the two surfaces would appear if greatly magnified. Notice the "peaks" and "valleys." In order to separate and move the surfaces apart or over each other without wearing off the asperities, or peaks (thereby eliminating the destructive friction forces), the introduction of a fluid film is required (see Fig.

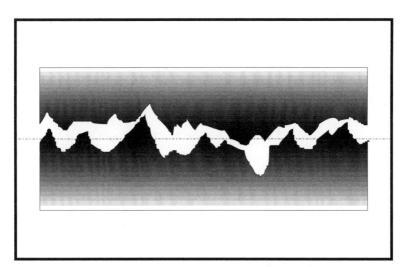

Fig. 2.2a. Metal surfaces greatly magnified under a microscope with no lubrication.

2.2b). The fluid film is the lubricant which separates the moving parts, allowing the friction to occur between the molecular planes of the lubricant instead of the metal surfaces. This action is called *"shearing."* A simple analogy would be to press hard on a deck of cards and move your hand forward. The cards act in much the same way as a lubricant, each card moving or "shearing" at a different rate, absorbing the generated friction, and allowing the hand to

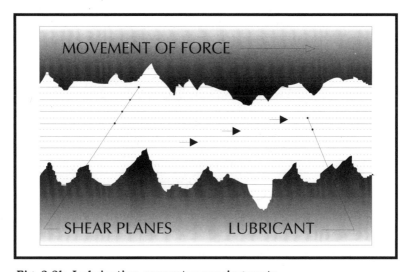

Fig. 2.2b. Lubrication separates moving parts.

Fig. 2.2c. Bearing surface at 2000× magnification.

(Courtesy Optimol Lubricants.)

Friction

causes

Heat

causes

Expansion
(High points come in contact and break off.)

causes

Wear
(Wear particles become tiny cutting tools that cause chatter, vibration, and more heat.)

To Combat Friction, it is Essential to

LUBRICATE!
Using the *right* amount, in the *right* place, at the *right* time.

Fig. 2.2d. Cause and effect of friction.

move freely across the table. Try it—it works! Fig. 2.2c shows an actual surface which has been magnified 2600 times. Notice the large ridges—these represent the actual peaks and valleys. Without lubrication to keep them separated, the moving surfaces begin to heat up, causing expansion to occur. Expansion then allows the opposing asperities to contact and collide with each other, attempting to retard the motion and causing the asperities to break away. These tiny points then have the ability to cause more friction and vibration, thereby setting up a destructive cycle. This direct cause-and-effect relationship of frictional force is shown in Fig. 2.2d.

The force of friction reveals itself in three different ways:

1. sliding friction

2. rolling friction

3. combination friction.

2.3 Sliding Friction

Sliding friction is common where any two plain surfaces move over one another. A common everyday example of sliding friction would be found within the internal combustion engine (see Fig.

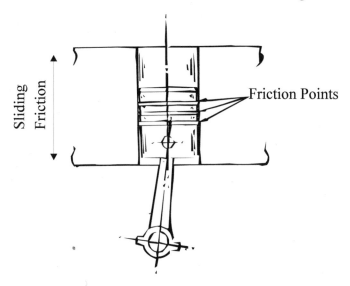

Fig. 2.3a. Engine piston cutaway. An example of reciprocating sliding friction—an internal combustion engine piston.

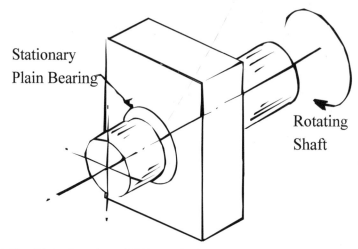

Stationary
Plain Bearing

Rotating
Shaft

Fig. 2.3b. Plain bearing. Sliding friction—rotational, shaft in a plain sleeve bearing.

2.3a). A piston moves up and down within a stationary cylinder. The piston rings constantly slide or rub against the cylinder walls creating sliding friction. Another example would be plain bearings or sleeve bearings (see Fig.2.3b) where a journal rotates within a sleeve. The rotation movement of the shaft against the sleeve would be described as sliding friction.

2.4 Rolling Friction

An important example of rolling friction is found in all rolling element bearings (see Fig. 2.4a), i.e., ball bearings, roller bearings,

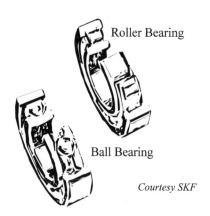

Roller Bearing

Ball Bearing

Courtesy SKF

Fig. 2.4a. Ball bearing and roller bearing.

Fig. 35.—*Discoloration and softening of metal caused by inadequate lubrication and excessive heat.*

Fig. 36.—*Glazing caused by inadequate lubrication.*

Fig. 34—a,b,c,d. *Progressive stages of spalling caused by inadequate lubrication.*

Fig. 37.—*Effect of rollers pulling metal from the bearing raceway.*

Fig 2.4b. Effects of poor lubrication on bearing surfaces.

(Courtesy SKF Bearing.)

needle bearings, taper roller bearings, etc. These types of bearings have often been termed "frictionless bearings." This is due to the small "point" contact nature of the ball rolling on the inner and outer races. But minor surface imperfections and dirt infiltration will cause a degree of friction that needs to be addressed.

Fig. 2.4b shows the effects of inadequate lubrication on rolling element bearing surfaces (courtesy of SKF Bearings).

2.5 Combination Friction (Sliding and Rolling)

The third type of friction is a unique combination of sliding and rolling friction occurring simultaneously. This combination friction is most often found in meshing gear sets (see Fig. 2.5.). When the teeth first come into contact and then disengage, sliding friction occurs. In between this happening, the opposing teeth pitch surfaces meet, causing rolling friction. Certain types of gear sets (e.g., hypoid gears, worm gear sets, etc.) produce a much higher degree of sliding friction.

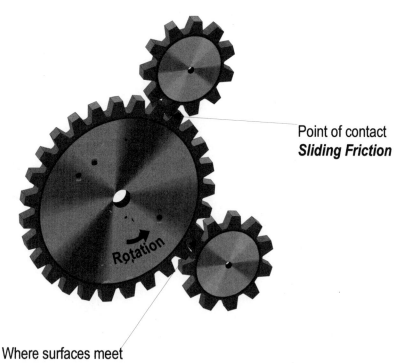

Point of contact
Sliding Friction

Rotation

Where surfaces meet
Rolling Friction

Fig. 2.5. Gears meshing results in combination friction.

2.6 Defining Lubrication

Lubricate *(Webster's Dictionary Definition):* To make smooth or slippery, to diminish friction by applying a lubricant.

(To ensure that two surfaces in relative motion do not come into contact.)

Lubricant *(Webster's Dictionary Definition):* A substance, e.g., grease, oil, soap, etc., that when introduced between solid surfaces which move over one another, reduces resistance to movement, heat production and wear by forming a fluid film between the surfaces.

2.7 Basic Functions of a Lubricant

In order to lubricate successfully, a basic understanding of a lubricant's function is necessary. A lubricant, whether it is oil or grease, can at any time perform up to six basic functions simultaneously. These functions are to:

1. reduce friction

2. reduce wear

3. absorb shock

4. reduce temperature

5. minimize corrosion

6. seal out contaminants.

REDUCE FRICTION

Reducing friction is the primary function of a lubricant. It does this by separating the two contact surfaces and allowing them to move over one another via the lubricant's viscous shear planes (see Fig. 2.2b).

REDUCE WEAR

With the introduction of a lubricant, we no longer have the asperities of the moving surfaces in collision: by virtue of this fact, the wear element particulates are significantly reduced and component life is extended (see Fig. 2.2d).

ABSORB SHOCK

Shock absorption is particularly significant within gear meshes. When meshing gears are not lubricated, the mating teeth set up shock waves. This is heard as a "chattering" sound. This type of shock often results in gear teeth fracture. Introduction of a lubricant reduces the chattering, acting as a "shock absorber." When a gearbox is properly lubricated, it runs more smoothly and relatively noise free.

REDUCE TEMPERATURE

A major benefit of reduced friction is reduction in operating temperature. Caution must be observed in the overall assessment here, because excessive lubricant may cause fluid friction which may in turn raise the temperature. This is explained further in Chapter 4.

Assuming that the correct quantities of lubricant are used, lubricants can be an excellent dissipator of heat, especially in recirculative oil (or splash oil) systems where the oil is passed over the moving part—where it not only lubricates, but absorbs the heat and returns to the reservoir where it cools before recommencing the cycle. (Sometimes it is necessary to pump the lubricant through an oil cooler, which will allow for a smaller reservoir.)

MINIMIZE CORROSION

If a lubrication barrier were not in place, moisture in the air would eventually cause oxidation, leading to corrosion. Lubricants cling to the element surface, providing a barrier against moisture. Again, the choice of lubricant type is critical because some lubricants act as a catalyst, trapping moisture droplets and holding them at the element surface, thereby increasing the oxidation and corrosion process!

SEAL OUT CONTAMINANTS

Basically stated, lubrication keeps the dirt contaminants out. Dirt is evident in all aspects of surface contact. It may show up as metallic wear particles or silicon particulates. The lubricant's job is to "flush" these contaminants out of the bearing surfaces so that they may be wiped away, as in the case of grease, or caught in a filtration medium (oil). Generally, the lubricant will also act as a seal against outside dirt ingestion.

2.8 Lubrication Film

As shown in Fig 2.2b, a lubrication film separates moving parts, allowing them to slide over one another. In order to understand exactly how a lubricant performs, it is important to understand the three basic lubrication film conditions that exist.

1. Full film $\begin{cases} \text{Hydrodynamic (HDL)} \\ \text{Hydrostatic (HSL)} \end{cases}$

2. Elastohydrodynamic (EHD)

3. Boundary layer.

Each condition relates to a film thickness present at the contact area. Film thickness has a direct effect on the degree of lubrication protection. This is commonly termed as the "F" lubrication factor.

FULL FILM

Full film is the ideal or preferred condition for sliding friction. Full film denotes the presence of enough lubricant to ensure complete separation of the moving surfaces. Earlier in this chapter, reference was made as to how a lubricant behaves in much the same way as a deck of cards, with the lubricant sliding along shear planes. In the case of a full film condition, the lubricant is analogous to a full deck of cards. The moving surfaces are fully apart, separated by a continuous fluid film. Full film lubrication is divided into two categories:

1. hydrodynamic full film (HDL)

2. hydrostatic full film (HSL).

If a journal is at rest within a plain bearing, a metal-to-metal condition prevails. As the journal rotates, it squeezes the lubricant between itself and the stationary bearing, forming a pressurized film. As the rotation increases to full speed, the journal is lifted

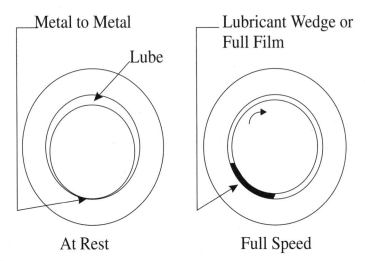

Fig. 2.8a. Hydrodynamic full film lubrication in a journal-type bearing.

away from the bearing surface on a wedge shaped film of oil, caus-ing no appreciable surface deformation to either the journal or the bearing. This is the basis of *hydrodynamic full film* lubrication (see Fig. 2.8a).

Through rotational forces, hydrodynamic lubrication produces its own pressure area which separates the moving elements. Hydro-static full film lubrication relies on an external device to produce its pressure lift and separation of moving elements. The device is usu-

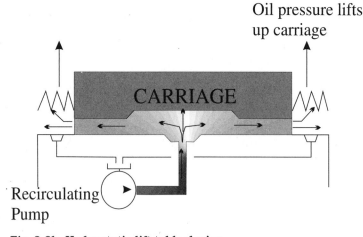

Fig. 2.8b. Hydrostatic lift table device.

ally a pump which develops enough pressure to support and float the load on a full film of oil. A usual application of a hydrostatic lift system would occur on a gear grinding machine, where the carriage holding the grinding wheel "floats" on a hydrostatic oil film (see Fig.2.8b).

ELASTOHYDRODYNAMIC FILM (EHD)

The elastohydrodynamic (EHD) film thickness is dependent upon its operating conditions, i.e., surface speeds, loads, lubricant viscosity, and the pressure–viscosity relationship. EHD is really a hydrodynamic film formed by applied pressure or load. It is predominantly found in rolling element bearings. The load on a roller in a roller bearing, for instance, causes it to move toward the stationary plate, or raceway in this case. This load causes a pressure area that elastically deforms, or "flattens out" (see Fig. 2.8c). This is termed the "Hertzian" contact area. In this area, the lubricant is placed between the rotating and stationary members. As the pressure increases, the lubricant is squeezed and begins to flow out be-

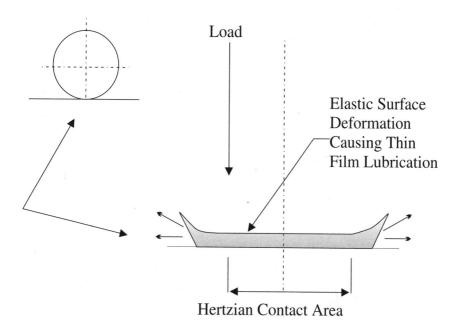

Fig. 2.8c. Elastohydrodynamic film in a rolling element.

tween the two members. This pressure can go as high as 200,000 psi and can squeeze the lubricant into a thin film. The resultant squeezing of the lubricant causes the lubricant's viscosity to increase, thereby allowing an "antifriction" state to be set up with an extremely thin film of lubricant. At this point, the fluid film acts as a solid and enables the ball or roller to "roll" cleanly and effectively against the inner and outer races, while at the same time receiving effective lubrication. Again, selection of the correct lubricant is critical in this type of bearing application.

BOUNDARY LAYER FILM

This is the final condition that occurs before component failure. It is usually the result of an insufficient lubricant supply, and/or the incorrect choice of lubricant which is incapable of supporting the external loads placed upon it.

Boundary layer (see Fig. 2.8d) is sometimes referred to as thin film lubrication. Although there is lubrication present, there is not enough to fill the element surface's valleys, thereby allowing metal-to-metal sacrificial contact to occur.

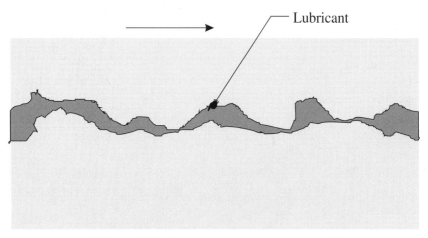

Lubricant

Although lubricant exists - surface separation
is not 100%, and asperity collision is inevitable.

Fig. 2.8d. Boundary layer.

2.9 The Three Generations of Lubrication

Since lubricants were first developed, technology has advanced lubricant manufacturing through three distinctive phases, or generations as they are often called. All three generations of lubricant are available and in use today. Each type of lubricant has the similar objective of performing constant hydrodynamic lubrication. How each generation of lubricant goes about its objective makes each uniquely different. The three lubricant generations are:

1. chemical wear lubricants (CWL)

2. solid lubricants (SL)

3. plastic deforming lubricants (PDL).

FIRST-GENERATION CHEMICAL WEAR LUBRICANTS

Active Sacrificial-Type Lubricants: Chemical wear lubricants (CWL) are often identified as the common everyday automotive and industrial lubricants, and are the least expensive to purchase. These conventional lubricants have been on the market since the 1920's and are marketed widely by all the major oil companies.

To prevent seizure, these types of lubricants are known as chemically aggressive. As the lubricant coats the metal surfaces, soft layers of metallic salts (sulfides and phosphides) are allowed to form at the contact surfaces. As the two surfaces slide over one another, alternating load cycles allow these soft surfaces to collide. When unit loadings exceed the sulfur phosphide film, the film ruptures and metal-to-metal contact occurs. This "weld and break" action results in high lubricant temperatures and the subsequent release of small metal particulates. The high temperature causes thermal degradation of the lubricant, with the metal particulates acting as catalysts, speeding oxidation, and resulting in shortened lubricant life. See Fig. 2.9a.

SECOND-GENERATION SOLID LUBRICANTS

Passive Semisacrificial Lubricants: The second generation of lubricants was introduced in the 1940's. This generation includes sol-

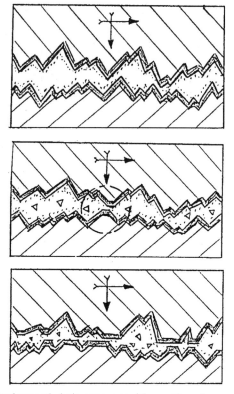

Fig. 2.9a. Chemical wear lubricants provide surface improvement through abrasion. *Top:* Surface at startup. *Middle:* Wearing in. *Bottom:* Finished surface.

ids such as graphite, molybdenum disulfide (MoS_2), boron nitride, PTFE (Teflon), titanium dioxide, etc. The lubricants are usually manufactured with a chemical wear base oil and a solid lubricant additive. They react differently than first-generation chemical wear lubricants in that a protective layer of solids is deposited in the valleys, coating the surfaces, thereby increasing the load bearing surface. As the two peaks pass, the solid additive flake shears, allowing the two deformed peaks to pass with limited damage.

These solid additives provide a smoother surface area through a less sacrificial process. Because of the size, cost, and nature of the solid additives (it is very difficult to attain higher than 4% volume of solids), we find that it is statistically impossible to continually have a solid additive flake between all clashing peaks; therefore, just as

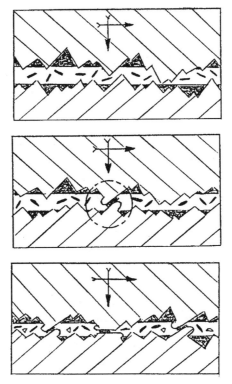

Fig. 2.9b. Solid lubricants provide surface improvement through buildup. *Top:* Solid lubricant startup. *Middle:* Wearing in. *Bottom:* Finished surface.

in chemical wear lubricants, "weld and break" occurs and metal particulates are released as before. See Fig. 2.9b.

Special Note: A first-generation lubricant will *not* become a second-generation lubricant by adding an after-market "solid" additive.

THIRD-GENERATION PLASTIC DEFORMING (PD) LUBRICANTS

Nonsacrificial Lubricants: Introduced in the 1970's, plastic deforming (PD) lubricants represent the latest technology in lubricants and additives.

When the peaks collide, a high load zone is set up (as described earlier); this creates temperatures of up to 1100°C. Under these conditions, the special PD liquid metallic additives penetrate the

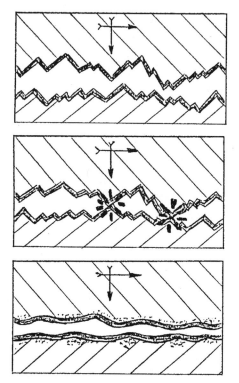

Fig. 2.9c. Plastic deforming lubricants provide surface improvement through metal restructuring. *Top:* PD startup. *Middle:* Wearing in. *Bottom:* Finished surface.

working surface and allow the surface to soften and deform without sacrificial loss. Repeated deformation restructures the complete surface to an almost flat state (over 70% flatness of theoretical maximum) and the metal returns to its original density.

When two machined, unlubricated surfaces are placed together, the actual surface contact is less than 10%. First-generation lubricants will allow sacrificial wear to occur and bring the flatness to 20% of the theoretical maximum. Second-generation lubricants bring the figure up to 40% of the theoretical maximum. However, third-generation lubricants accelerate the surface flatness to 70% of the theoretical maximum with virtually no wear. This results in less friction, lower lubricant requirements to reach full film state, lower running temperatures, and extended lubricant changeouts. This PD lubricant is often termed a "surface engineering" product.

It can be used to take a pitted surface and almost completely renew it. See Fig. 2.9c.

COMPARISON OF SURFACES

The photographic enlargements in Fig. 2.9d illustrate the finished state of metal after break in has occurred with all the different generation types of lubricant.

Second- and third-generation lubricants are considerably more expensive than first-generation lubricants, and can be mineral or synthetic based. In most applications, first-generation lubricants will offer more than adequate protection. Return on investment (ROI) figures will differ greatly depending on the type of equipment

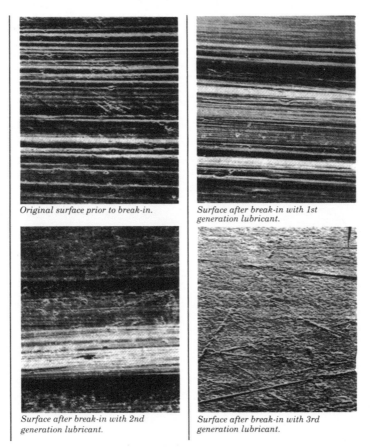

Original surface prior to break-in.

Surface after break-in with 1st generation lubricant.

Surface after break-in with 2nd generation lubricant.

Surface after break-in with 3rd generation lubricant.

Fig. 2.9d. Comparison of lubricated surfaces.

(Courtesy Interlube.)

and industry. A lubrication audit by a qualified consultant is advisable when determining the right kind of lubricants, delivery systems, and storage methods to use. Refer to the following Case Studies.

Review Questions

1. Define tribology.

2. In what four critical areas does tribology affect industry?

3. Define friction.

4. What is "shearing?"

5. Name three types of friction and give a working example of each.

6. Name the six basic functions of a lubricant.

7. Name three types of lubrication films—explain one.

8. Name the three generations of lubricants—explain one.

Case Study I

LOOKING FOR ENERGY SAVINGS THROUGH LUBRICATION ENGINEERING— FIRST-VERSUS THIRD-GENERATION LUBRICANTS

A hydro utility's energy conservation division recently performed an *"electrical cost reduction through better friction control"* study. The donor facility for this study was an automotive parts supplier stamping plant. This type of facility was chosen based on the degree of difficulty factor, i.e., if it could be proven that lubrication engineering reduced energy consumption on a stamping press (which is subjected to cyclical power demands; and which has rotary, linear, and arc movements all within one stroke of the press), it would have substantially proven the theory that lubrication engineering can reduce energy requirements for mechanical equipment. Two presses were chosen, as follows.

1. Press #1 was a continuous-operation 500-ton straight side mechanical press which utilized an automatic recirculative oil, series progressive-type lubrication system. The lubricant in use was a first-generation chemical wear EP 150 lube. The press was powered by a variable speed drive. No oil leakage was apparent.

2. Press #2 was a 600-ton fixed drive single stroke operated straight side press. The press was grease lubricated via an automatic series progressive lubrication system. The lubricant in use was a first-generation EP 1 grease.

Using an EL control energy analyzer, the Hydro Utility technician monitored the prelubricant changeout energy consumption figures over an accumulated period of two weeks. The average kW energy usage was noted. The lubricant delivery systems were then audited, cleaned, and recalibrated. Overlubrication was a factor in both presses—particularly in Press #2.

The lubricant was changed out and filters were replaced. Press #1 received a third-generation PD-type 150 weight oil, and Press #2 received a full line flush and was refilled with a third-generation PD #1 grease. The equipment was then run under the same load conditions (same manufactured part) and remonitored for the next 10-day span. *(N.B.: The third-generation lubricants used were mineral-based lubricants, **not** synthetic lubricants.)* When results were compared, it was clearly shown that effective lubrication **does** reduce electrical consumption considerably. The results showed the following reduction figures:

(oil) Press #1—Reduction in power consumption = 17.92%
(grease) Press #2—Reduction in power consumption = 8.92%

Additional benefits incurred were:

- extended oil change intervals (up to 4 times)
- enhanced load carrying capability
- enhanced antiwear
- noise reduction
- wear surface improvement.

Case Study II

EXAMPLE OF SAVINGS INCURRED THROUGH LUBRICATION ENGINEERING

As seen in the previous case study, the extended benefits of using premium lubricants are an added bonus to energy savings. In some cases, the additional benefits can return substantially more than just energy savings.

A recent study performed for a petrochemical company, which was using large gas compressors, showed compressor shutdowns for lubricant changeout that did not coincide with the general plantwide turnaround shutdown. This compressor downtime cost the company well over half a million dollars per occurrence, per compressor.

By changing out the compressor's first-generation lubricant to a second- or third-generation lubricant, the company is now able to take advantage of the second- and third-generation lubricants' extended changeout benefit. The lubricant can now be changed out during plant turnaround, which translates into millions of dollars of downtime savings over the next 5–10 years as well as energy consumption savings.

CHAPTER 3

Oil and Grease

3.1 Oil Classifications

Oils fall into the following three categories:

1. animal/vegetable

2. mineral

3. synthetic.

ANIMAL/VEGETABLE

Generally speaking, the animal/vegetable oils are not utilized for industrial bearing applications, due to acid formation after short periods of use. This acid is detrimental to the bearing surface performance. Animal/vegetable oils are generally reserved for cooking purposes.

MINERAL OILS

The majority of bearing surface lubricating oil is derived from refined crude petroleum. The refining method is either by the *solvent refining process* or the *hydrotreating process.* Both methods consist of a series of processes designed to remove undesirable components such as aromatic hydrocarbons, acids, sulfur compounds, and wax. They also improve desirable properties such as viscosity index, pour point, and stability. After refining, we end up with a base stock oil to which additives are blended to enhance the lubricant's specification.

Example: Additives are added to reduce oxidation, reduce foaming, and enhance antiweld or EP (extreme pressure) properties. There are antirusting agents and performance enhancers such as molybdenum disulfide, PTFE, graphite, and plastic deforming agents (see "Oil Additives," p. 47).

SYNTHETIC OILS

Synthetics are man-made lubricants designed to work under conditions where normal petroleum base oils would find limitations. Because synthetics are man made, they are usually more consistent and uniform in structure than petroleum base stocks.

Advantages

· Because no waxes are present, synthetics can be used at very low temperatures.

· Synthetics can also be used at much higher temperatures. Petroleum-based oils have a temperature application limit of approximately 600°F (320°C), after which the base oil decomposes and cokes.

· At higher temperatures, sludging and acid buildup are less apparent with synthetics, due to their better oxidation stability.

· Synthetics have a more stable Viscosity Index (V.I.) and can be relied upon to be a more stable lubricant over different temperature values (see Viscosity Index, later in this chapter).

· Synthetics are instrumental in extending lubricant changeout intervals (oil changes).

· Synthetics allow the equipment to run cooler, thereby conserving energy.

Disadvantages

· Synthetics are much more expensive to purchase. Prices range from five times mineral base lubricant cost, and up.

· Many synthetics are not compatible with certain sealing materials and may require replacement of seals, hoses, and paint.

Property	Mineral Oil	Polyalphaolefin	Dibasic Acid Ester (Diesters)	Polyglycol (PAG)	Silicone
Viscosity Characteristics (Temperature)	Moderate	Good	Excellent	Good	Excellent
V.I.	Moderate	Very Good	Good	Very Good	Very Good
Low Temperature Pour Point	Good	Very Good	Very Good	Good	Good
Oxidation Stability	Moderate	Very Good	Good	Very Good	Very Good
Volatility	Moderate	Good	Very Good	Good	Good
Lubricating Properties	Good	Good	Very Good	Good	Moderate
Mineral Oil Compatibility	—	Excellent	Good	Poor	Poor
Cost	Low	Medium	Medium	Medium	High

Fig. 3.1. Property comparisons of lubricants. These property comparisons are for guide purposes only; for specific data, consult the lubricant manufacturer.

There are two basic categories of synthetic lubricants.

1. Synthesized Hydrocarbons

· Polyalphaolefins (PAO's)

· Polybutenes

2. Organic Esters

· Polyol Esters

· Dibasic Acid Esters (Diesters)

There are literally hundreds of available synthetic oils; each synthetic stock type acts in a specific way and is designed for a specific purpose. Speak with reputable dealers when deciding to use a synthetic lubricant. Its use should be based upon operational problem solving that can justify and offset the much higher costs. Fig. 3.1 shows the property comparisons of a typical mineral oil compared to a variety of synthetic oils (courtesy STLE).

Rule of Thumb: Do not mix synthetic lubricants with each other or with petroleum base oils. Serious component damage could occur. As with all changeout situations, consult with the lubricant manufacturer as to the proper procedures for flushing and filling.

3.2 Oil—A Comparison of Advantages and Disadvantages

As a lubricant for load bearing surfaces, oil possesses certain advantages and disadvantages. It is the preferred medium and will always have the advantage over grease. See Fig. 3.2.

3.3 Viscosity

Generally speaking, lubricating oils are very uniform in their characteristics. The primary area of investigation when choosing the correct lubricant should be the oil's viscosity.

Viscosity: a measure of a lubricant's resistance to flow.

The viscosity of a fluid is entirely dependent on its temperature, load, and physical state. Viscosity bears a direct relation to the lu-

ADVANTAGES	DISADVANTAGES
Easy to apply	Hard to control
Oil lube systems are usually less expensive	Could contaminate product and machinery
Excellent cleaning and flushing characteristics	Correct viscosity selection is critical in retaining good performance
Can be used in recirculative systems	Susceptible to temperature variations
Within systems, generally more stable as a lubricant than grease	Leakage
With correct application, no limit to machine speed	

Fig. 3.2. Oil comparison chart.

bricant's film strength and ability to keep moving parts separated. See Fig. 3.3a.

The importance of proper viscosity selection cannot be stressed enough. If an application involves high speeds, low loads, and low

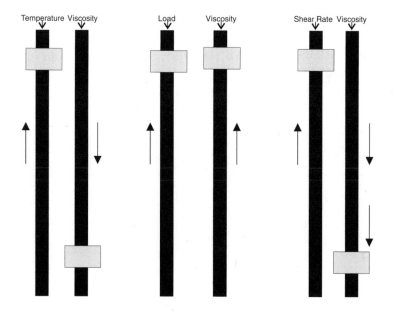

Fig. 3.3a. Viscosity behavior.

Case Study III

SYNTHETIC VERSUS MINERAL-BASED LUBRICANTS IN ADVERSE SITUATION PRODUCE SIGNIFICANT SAVINGS

For the majority of situations, chemical wear first-generation lubricants perform adequately; for adverse conditions, such as extreme temperatures, the use of premium lubricants really *pays* off.

A hydro utility study on energy reduction through lubrication engineering studied the use of synthetics in air compressors. The compressed air unit in the study was a 150-hp screw-type compressor that was recently rebuilt and utilized a standard 32 weight lube oil. The conveyor was checked by a compressed air audit company who gave it a clean bill of health. The power was monitored under varied load conditions and its power consumption noted. Because compressors run extremely hot, synthetic lubricants are well suited to this type of application. Not only do synthetics allow the press to run cooler, they also "desludge" or clean the compressor and allow it to run more efficiently, thereby reducing power consumption. Synthetics also allow for extended changeout intervals. The standard lubricant was replaced with a 32 weight fatty acid ester-type synthetic (second-generation) lubricant. An energy reduction of 7.3% was gained by changing the lubricant, and the changeover also resulted in a cooler running compressor.

temperatures, then a low viscosity or "thin" lubricant is adequate. Conversely, if low speeds, high loads, and high temperatures were the given, then high viscosity (or "thick") lubricant should be chosen.

It is necessary to ensure that the viscosity is high enough to provide a continuous oil film in the contact area, but not too high so as to create fluid friction due to *viscous* shear. (This is a condition set up when the oil is too thick for the application. This causes the shear planes to be "dragged" over one another, as opposed to their normal sliding motion. This in turn causes fluid friction—a common name for this is *churning*.) The ambient and operating temperatures will also affect the viscosity. On a cold morning, the engine of a car turns over much more slowly on startup. This is due to the cold temperature (ambient) "thickening" the lubricant, allowing the viscosity to become much higher. When the engine warms up, the oil "thins," reverting to its design viscosity at operating temperature. Automobile lubricants (multigrade oils) are designed to work in cold and hot conditions. The majority of industrial lubricants are single grade oils and are therefore much more reliant on a stable ambient and operating temperature. At present, there is no worldwide uniform standard measuring system for viscosity. Fig. 3.3b compares various popular measures in use today. (Also see the Viscosity Limitation Guide below.)

Viscosity Limitation Guide
(For Reference Purposes Only)
Maximum Viscosity (at Startup) in Centistokes:

22,000	—Maximum pouring viscosity
11,000	—Maximum viscosity for splash of bath lubrication
8,600	—Maximum pumpability viscosity for gear and piston pumps
2,200	—Upper limit for automatic oil lubrication pumps
2,200	—Upper limit for oil constituent of dispensable grease
1,000	—Ring of roller element bearings
860	—Maximum limit for hydraulic vane pumps
220	—Oil mist systems

SUS @ 100F	ISO	AGMA	SAE	GEAR	SUS @ 210F	CST @ 40C
32	2	-	-	-	-	2
60	10	-	-	-	-	10
105	22	-	-	-	-	22
170	32	-	10W	75W	44	32
240	46	1	20	-	48	46
350	68	2	20	80W	55	66
480	100	3	30	-	63	100
750	150	4	40	85W	75	150
1050	220	5	50	90	95	225
1650	320	6	60	90	115	320
2500	460	7	70	140	145	460
3500	680	8	-	140	180	680
5000	1000	8A	-	250	250	1000
8000	1500	-	-	250	-	1500

Fig. 3.3b. Lubricant viscosity rating and comparison chart (ratings are approximated).

SSU = Saybolt Second Universal
ISO = International Standards Organization
AGMA = American Gear Manufacturers Association
SAE = Society of Automotive Engineers
CST = Centistokes

220 —Hydraulic piston pumps (to prevent wear)
54 —Hydraulic systems at operating temperature

Minimum Viscosity in Centistokes:
33 —Gear lubrication
30 —Gear pumps
21 —Spherical roller bearings
13 —Other rolling element bearings
13 —Plain bearings
13 —Hydraulic systems
4 —Minimum viscosity to support dynamic load

Case Study IV

THE EFFECTS OF ADVERSE TEMPERATURE ON VISCOSITY

This study involved a tire manufacturing plant. A tire press employed a two line parallel lubrication delivery system. The pumping unit was over 20 years old but had performed well over the years. The original grease in use was an NLGI (National Lubricating Grease Institute) #1 EP-type grease. This was changed to a #1½ grease (there is no 1½ rated grease—the custom blender of this grease marketed it as a #1½ grease; when tested, this grease would actually be classified as an NLGI #2 grease), and immediate problems occurred when the lubricant delivery pump failed to pump. The company changed out the pump on the premise that its causal factor for failure was old age, but the problem continued to occur.

The grease was tested and found to actually be NLGI #2 grease (this should still have had good pumpability properties). The new pump operated normally when disconnected from the system. The delivery system also checked out to be in good working order. Further investigation found a series of broken windows near the pump station. Winter time was looming and the temperature had dipped drastically. The ambient temperature in the pump station room varied throughout the day, which explained why the pump worked only some of the time. Coupled with a heavier grease and intermittent low ambient temperatures, the oil's viscosity in the grease had thickened to the point where it was difficult to pump through the automatic grease system, and the grease pump would stall.

The problem was solved with a blanket-style heater which was wrapped around the lubricant reservoir and operated with an on–off timer. This solution enabled the old pump to be recommissioned in its old role, and the new one to be utilized elsewhere in the plant.

Technical data

	Unit	22	100	150	220	320	460	680	1000	3000
Color	-	brown	brown	brown	brown	brown	brown	brown	brown	brown
ISO viscosity group AGMA	-	22 -	100 3 EP	150 4 EP	220 5 EP	320 6 EP	460 -	680 -	1000 -	3000 -
Density at +15°C/+59°F	g/cm³	0.890	0.901	0.904	0.910	0.917	0.920	0.930	0.927	0.918
Dyn.visc. at +40°C/+104°F	mPa.s	19.5	89.2	142	198.3	293.4	422.3	623	927	2750
Kin.visc. at +40°C/+104°F	mm²/s	21.9	101	157	218	320	459	670	1000	3000
Kin.visc. at +100°C/+212°F	mm²/s	4.35	11.4	15.0	18.8	24.3	30.6	37.8	50.0	122.0
Viscosity index	VI	106	99	95	97	97	96	94	95	117
Flash point	°C °F	165 329	224 435.2	226 438.8	232 449.6	236 456.8	236 456.8	238 460.4	260 500	260 500
Pour point	°C °F	-34 -29.2	-24 -11.2	-18 -0.4	-16 +3.2	-15 +5.0	-12 +10.4	-9 +15.8	-9 +15.8	0 +32
Copper corrosion 48h/+100°C/+212°F	-	A 1	A 1	A 1	A 1	A 1	A 1	A 1	A 1	A 1
SRV® test run Test mode 5ae (2h, 300N, +50°C/ +122°F/ball/area) Friction coefficient	μ (min)	0.035	0.030	0.030	0.030	0.025	0.025	0.025	0.025	0.025

Technical data, *continued*

	Unit	22	100	150	220	320	460	680	1000	3000
Friction coefficient (running-in period)	μ (max)	0.130	0.130	0.130	0.130	0.130	0.130	0.130	0.130	0.130
Wear: a) ball/scar φ	mm	0.85	0.85	0.85	0.85	0.85	0.85	0.85	0.85	0.85
b) profile depth Pt	μm	0.8	0.8	0.8	0.8	0.8	0.8	0.8	0.8	0.8
Lubrimeter test run Test mode M2 abrasion	mg	0.062	0.062	0.062	0.062	0.062	0.062	0.062	0.062	0.062
Friction coefficient μ(5')	-	0.042	0.040	0.040	0.040	0.035	0.035	0.035	0.035	0.035
Friction coefficient μ(60')	-	0.040	0.038	0.038	0.038	0.034	0.034	0.034	0.034	0.034
Wear scar depth	μm	0.5	0.5	0.5	0.5	0.5	0.5	0.5	0.5	0.5
Load carrying capacity	%	98	98	98	98	98	98	98	98	98
FZG Boundary lubrication A/8, 3/90	Damage load stage	>12	>12	>12	>12	>12	>12	>12	>12	>12
Specific weight change	mg/kWh	<0.27	<0.27	<0.27	<0.27	<0.27	<0.27	<0.27	<0.27	<0.27
Seal compatibility Volume change	%	+1	-2.5	-2.5	-2.5	-2.5	-2.5	-2.5	-2.5	-2.5
Shore hardness	Shore	0	+3	+3	+3	+3	+3	+3	+3	+3

These technical data are based on average test results. Minor deviations may occur from case to case.
Further product information is available on request.

Fig. 3.3c. Typical lubricating oil specification.

3.4 Viscosity Index

Viscosity Index (V.I.) is the rate of change in viscosity due to the change in temperature. The higher the V.I. of an oil, the less the tendency for its viscosity or thickness to change with changes of temperature. Synthetic oils tend to have a better Viscosity Index than petroleum-based stocks. If two lubricating oils were under consideration, all things being equal, the lubricant with the higher V.I. would be the lubricant of choice. See Fig. 3.4.

When viscosity at two temperatures is known (100°F and 210°F), locate as points on the lines. A straight edge is then laid on these points and intersects the Viscosity Index line. The point of intersection gives the V.I., which is read directly.

If viscosity at one temperature and a desired V.I. are known, locating their respective points with a straight edge enables determination of viscosity at the second temperature necessary to achieve the required Viscosity Index.

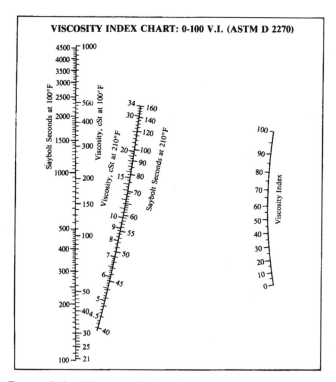

Fig. 3.4. Determining Viscosity Index (V.I.)—ASTM D-2270.

Case Study V

THE EFFECTS OF ADVERSE TEMPERATURE ON VISCOSITY

When a commercial fishing vessel works in the Northern oceans, it works under very demanding conditions. If that vessel incorporates a factory that flash-freezes fish, it requires an overhead conveyor system. The conveyor must run through the freezer and outside into the elements under adverse conditions.

One such company had an automatic conveyor lubricator that "shot" lubricant through dynamic injectors into the chain link and ball race areas of the conveyor. When in arctic areas, the conveyor lubricant thickened to the point that prohibited proper operation of the lubricator; the viscosity was thickening due to the cold conditions. To solve the problem, a fixed temperature automotive block heater was installed in the reservoir and turned on when required. The company no longer had a viscosity problem as the block heater enabled the lubricant to be kept at a constant temperature.

Rule of Thumb: The service life of a petroleum-based oil is specified as

30 years @ 85°F or 30°C
15 years @ 104°F or 40°C
3 months @ 212°F or 100°C

For each 18°F (or 7.8°C) increase, the lubricants effective life is halved. These figures are oil industry figures for guideline purposes. Other service factors should be considered when using this service life guideline.

3.5 Oil Characteristics

The most important characteristics of oil, as reviewed in Sections 3.3 and 3.4, are its viscosity and viscosity index. There are, however, other characteristics that must be taken into account when choosing the correct oil for the job.

POUR POINT

This is the point at which the oil solidifies (important for low temperature applications) and is no longer pourable.

FLASH POINT

At this temperature, the lubricant vapor will flash ignite.

FIRE POINT

This is a higher temperature than the flash point, and is the temperature at which the lubricant will catch fire and stay alight.

CORROSION OR RUST

This is the lubricant's ability to deter rusting or corrosion.

DEMULSIBILITY

This is the lubricant's ability to separate from water.

ANTIFOAMING

The lubricant has the ability to settle out foaming or entrapped air bubbles (especially critical in hydraulic oils); this will also deter oxidization.

Parameter	DIN	ASTM	IP	ISO
Viscosity	51 562	D445	71	
Viscosity Index	DIN/ISO 2909	D2270	226	2909
Pour Point	51 597	D97	15	3015
Flash Point (Open)	51 376	D92	36	2592
Flash Point (Closed)	51 758	D93	34	2719
Neutralization Number	51 558	D974	139	DP6618
Water Content	51 777	D4377	356	
Demulsibility	51 599	D1401	—	
Rust	51 585	D665	135	
Foaming	51 566	D892	146	DP6247
Air Release	51 381		313	
Four Ball	51 350	D2783	239	
Timken	—	D2782	240	
FZG	51 354 Pt. +2	—	334	

Fig. 3.6. Governing bodies—standards.

  IP = Institute of Petroleum (U.K.)
  ASTM = American Society for Testing and Materials (U.S.A.)
  DIN = Deutsche Industrie Norm
  ISO = International Standard Organization (Europe)

3.6 Oil Standards

Throughout the world, tests are performed under different standards organizations (see 7.2 ASTM test for lubricants). Fig. 3.6 shows a comparison of basic oil standard tests. (Also see the Specification Chart below.)

3.7 Oil Additives

Additives are blended with the petroleum or synthetic base stocks to strengthen or modify the characteristics of the lubricant. This allows the modified lubricant to meet much higher demands and specification requirements. The most common additives are the following.

DETERGENTS/DISPERSANTS

These materials chemically react with the oxidation products and enable the oil to suspend dirt particles and debris which can be taken out through filtration. They also help to neutralize acid buildup.

EP ADDITIVE

Extreme pressure (EP) additives improve the lubricant's performance where high local pressures exist.

ANTIWEAR AGENTS

These materials enable us to develop second- and third-generation lubricants, and are primarily used to prevent metal-to-metal contact on heavily loaded surfaces.

ANTIOXIDANTS

Oxidation inhibitors prevent oxygen from attacking the base oil, thereby stabilizing the lubricant's viscosity; this in turn prevents organic acid buildups.

VISCOSITY IMPROVERS

These additives are used to prevent the oil from "thinning" at elevated temperatures.

POUR POINT DEPRESSANT

These additives help to keep the oil "thin" at lower temperatures by preventing the formation of wax crystals.

It is important to realize that the base oil stock does not wear out. When a lubricant fails to give good service, it is the depletion of the additive package that is usually the cause of the degradation.

3.8 Changeout of Brands or Types of Oil

When changing out oil lubricants, the methods and procedures are entirely dependent upon the type of lubrication delivery system,

Case Study VI

ALL LUBRICANTS ARE NOT EQUAL

This study was done in a large automotive assembly plant which was experiencing multiple air tool and hydraulic valve failures. Through the use of the work order management system, the investigation identified that the failures had started gradually and had intensified into a virtual line shutdown situation over a period of months (a line shutdown is measured in thousands of dollars per minute).

Wear Particle Analysis samples were taken at the lubrication delivery points where the mechanical failures were occurring, and at the plant lubricant storage facility where the clean lubricant was stored. Virgin stock samples were also obtained directly from the lubricant manufacturers. The results found all three samples showing markedly different results.

Further investigation found that lubricants were purchased from a variety of manufacturers based on viscosity specifications, e.g., hydraulic 32 weight oil. These virgin stocks were then used to "top up" the storage containers in the lubricant storage room. The topped-up lubricants would then be transferred to the lubricant dispensing equipment. The transfers would take place with the *same* transfer pump for *all* lubricants. Also, there were only two funnels for all lubricants. These funnels were not thoroughly cleaned between lubricant transfer use, thereby adding significantly to the contamination of the lubricants.

The problem was a compound one. Oils were being allowed to be mixed at the receiving stage as well as the transfer stages. Incompatibility of lubricants "stripped out" the additive package and neutralized other additives which rendered many of the lubricants ineffective. To solve the problem, a three-part solution was implemented.

1. The lubricant purchasing specification was rewritten. The new specification was highly specific in that it named the actual lubricants as opposed to identifying only the weight or class of lubricant.

2. All lubricant storage vessels in the lubricant storage area were labeled with the lubricant brand name and relevant information of the lubricant they contained.

3. New transfer pumps and funnels were purchased. Each individual lubricant stored was issued its own personal transfer pump and funnel. The pumps and funnels in turn were labeled with the lubricant brand name.

type of old lubricant, type of new lubricant, compatibility to paint and seals, and the need for system flushing. In order to effect a successful oil type or product brand change, it is recommended that the supplier of the new lubricant be charged with the responsibility of devising a suitable changeout procedure.

3.9 Grease—A Comparison of Advantages and Disadvantages

Fig 3.9 shows some of the advantages and disadvantages of using grease as a lubricant.

Advantages	Disadvantages
Stays where it is placed	Can cook in high temperature applications
Withstands heavy shock loads	Excessive amounts of dirt or grit will not settle out and could harm bearings
Has excellent seal forming characteristics: keeps out moisture and dirt	Heavier greases cause heavier machine power consumption
Generally requires less frequent applications	Dependent upon frictional heat to let the oil content get into the contact area and lubricate

Fig. 3.9. Grease comparison chart.

3.10 Grease Classification

Oil and grease are basically the same, with one important difference—*oil is fluid, flowing readily by itself;* grease is also an oil product, but is made semisolid by combining it with chemical soap. A good analogy for grease would be a sponge soaked in oil. The word "grease" itself is derived from the Latin word "crassus" which means

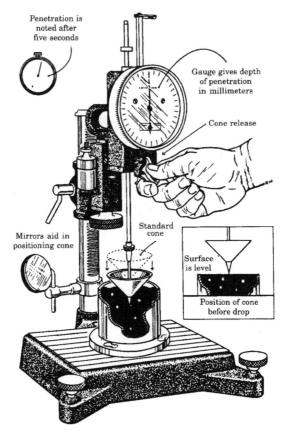

Penetration is
noted after
five seconds

Gauge gives depth
of penetration
in millimeters

Cone release

Mirrors aid in
positioning cone

Standard
cone

Surface
is level

Position of cone
before drop

Fig. 3.10a. ASTM D-217 cone test for grease.

fat. Grease relies on pressure to move it around; and although it is termed semisolid, it can range from very fluid to brick-hard.

Unlike oil, there is a single standard grading system for grease. This is the NLGI (National Lubricating Grease Institute) grading system, which identifies the grease's consistency. To test its consistency, a test called the ASTM D-217 cone test is used, whereby a special cone is dropped into the grease from a specified height, at a temperature of 77°F or 25°C, and its depth of penetration is measured in 0.1-mm increments (see Fig. 3.10a). This then corresponds to the NLGI numeric measurement. Most industrial applications utilize an NLGI #1 or #2 grade grease. See Fig. 3.10b.

Special Note: It should be noted that centralized grease lubrica-

Consistency NLGI Grade #	ASTM D-217 Cone Penetration (0.1 mm @ 77°F)	Appearance at Room Temperature
000	445–475	Very Fluid
00	400–430	Fluid
0	335–385	Semi-Fluid
1	310–340	Very Soft
2	265–295	Soft
3	220–250	Medium Hard
4	175–205	Hard
5	130–160	Very Hard
6	80–115	Block

Fig. 3.10b. NLGI grease classifications.

tion delivery systems are only rated for NLGI #1 and under. Check with the system supplier and lubricant supplier for compatibility.

It should be remembered that when choosing a grease, we are choosing an oil that has been thickened with a chemical soap, therefore viscosity rules still apply. In order to thicken grease, a process known as saponification is utilized.

SAPONIFICATION

Saponification is a process whereby fat or fatty acids react with an alkali to form a soap. The fat can be of animal, vegetable, or synthetic materials, with the alkali being a substance of basic properties. In the grease industry, saponification is considered to be the formation of a soap through the reaction of a fat with a metallic hydroxide. This metallic hydroxide is the component by which the grease is classified (e.g., aluminum, barium, calcium, lithium, sodium, etc.). Fig. 3.10c shows the different grease properties that are derived from each type of thickener. (Also see the Specification chart Fig. 3.10d on page 56.)

Special Note: Information is intended as a guide only. For more specific information, contact the lubricant vendor.

Type	Approx. Temp. Range	Cold Weather Pumpability	Water Resistance	Usual Appearance	***Note***
Lithium	350F 175C	Very Good	Very Good	Buttery	Most commonly used grease.
Lithium Complex	375F 190C	Very Good	Very Good	Buttery	
Calcium (Lime Soap)	230F 110C	Fair	Very Good	Buttery	
Sodium (Soda Soap)	250F 120C	Fair	Poor	Fibrous	
Calcium Complex	350F 175C	Fair	Good	Stringy	Can bleed and harden in centralized systems.
Barium	380F 193C	Poor	Excellent	Fibrous	
Aluminum Complex	350F 175C	Good	Excellent	Stringy	Can form gel at high temperature, increasing power consumption and reducing wear properties.
Bentones (Non Soap)	500F 260C	Good	Good	Buttery	Can bleed and harden in centralized systems. Can be highly flammable due to wicking.

Fig. 3.10c. Grease thickeners.

3.11 Grease Characteristics

This section explains the various unique characteristics of grease.

PUMPABILITY

Pumpability refers to the grease's ability to flow easily under pressure through a distribution system over a given temperature range.

OIL SEPARATION

Oil separation is often termed as "leeching" or "bleeding," where the oil separates from the soap. A partial action of oil separation is required for bearing lubrication, but excess action can leave a soap-rich fraction. This in turn may cause partial or full flow restriction of the grease line. Be aware that some greases separate under pressure and can present problems in centralized delivery systems.

SLUMPABILITY

Slumpability refers to the grease's ability to move within a reservoir or container, under gravity, and maintain a seal at the pump outlet. Utilization of a follower plate will minimize the degree of slumpability by exerting a slight pressure on the grease within the reservoir or container.

DROPPING POINT

This is the temperature at which the grease softens enough to drip.

COLOR

Greases are colored for the manufacturer's benefit only.

WATER RESISTANCE

This refers to a grease's ability to withstand water "wash out" and prevent rusting *if* the water and grease *do* mix in the bearing area.

Rule of Thumb: Always check with the grease lubricant manufacturer before using a grease in a centralized lubrication system.

Product Code		EP2 504-700	EP MDS 504-701
NGLI Grade		2	2
Color		Green	Black
Thickener		Lithium Complex	Lithium Complex
Appearance		Smooth, tacky	Smooth, tacky
Worked Penetration 60 strokes at 25°C	(ASTM D 217)	280	280
Estimated Max. High Operating Temp. °C[1]		170	170
Estimated Min. Low Dispensing Temp. °C[1] Grease Gun		−30	−30
Dropping Point °C	(ASTM D 2265)	262	262
Low Temperature Torque @ −40°C, N.m.	(ASTM 4693)	4.54	4.54
Mobility, @ −17.8°C g/min	(USS DM 43)	6.5	6.5

Mineral Oil	(ASTM D 445)		
Viscosity cSt at 40°C		175	175
Viscosity Index	(ASTM D 2270)	96	96
Bleed @ 25°C	(IP 121)	3.1	3.1
Wear, 4-Ball Scar, mm	(ASTM D 2266)	0.4	0.4
EP Tests			
4-Ball, Load Wear Index	(ASTM D 2596)	45	47
Timken OK Load, kg	(ASTM D 2509)	20.5	20.5
Water Spray Resistance % Loss	(ASTM D 4049)	54	54
Elastomer Compatibility	(ASTM D 4289)	pass	pass
Bomb Oxidation at 99°C	(ASTM D 942)		
Pressure Drop at 100 h, kPa		30	35
Corrosion Test, 48 hr.	(ASTM D 1743)		
@ 52°C, Rating Number		1	1

[1] May vary with design of grease gun or application

Fig. 3.10d. Typical lubricating grease specification.

	Lithium Complex	Lithium	Aluminum Complex	Calcium Complex	Barium	Sodium	Bentone	Silica Gel	Polyurea
Lithium Complex	—	✓	✗	✓	✗	✗	✗	✓	✓
Lithium	✓	—	✗	✓	✗	✗	✗	✓	✓
Aluminum Complex	✗	✗	—	✗	✗	✗	✗	✓	✗
Calcium Complex	✓	✓	✗	—	✓	✗	✗	✗	✗
Barium	✗	✗	✗	✓	—	✗	✗	✓	✗
Sodium	✗	✗	✗	✗	✗	—	✗	✗	✗
Bentone	✗	✗	✗	✗	✗	✗	—	✓	✗
Silica Gel	✓	✓	✓	✗	✓	✗	✓	—	✗
Polyurea	✓	✓	✗	✗	✗	✗	✗	✗	—

Fig. 3.12. Guide to grease compatability. This chart is an accepted general guide *only*. Compatability can only be accurately determined on an individual basis. When changing lubricants, contact the replacement vendor for compatability assurance and instructions for lubricant changeout.

3.12 Grease Compatibility

Fig. 3.12 shows how important it is *not* to mix greases inadvertently. This is for guide purposes only—always contact the manufacturer before changing grease types.

Although oil is the preferred medium for lubrication purposes, grease has a definite advantage for remote applications and on equipment where oil is hard to contain. The replenishment cycle is also less frequent (further explanation is found in Chapter 4).

3.13 Pumping Grease

In order to move grease to the application point, it needs to be pumped. Most delivery systems for oil utilize pumps that work in the 50 psi (3.4 bar)–250 psi (17.2 bar) range. Grease, on the other hand, requires a pump that will deliver pressure above 1500 psi (103 bar).

In the hands of an untrained operator, an ordinary grease gun can deliver a pressure of up to 15,000 psi (1030 bar)! A bearing seal, on the other hand, will rarely rate higher than 500 psi (35 bar). Once the seal on a bearing is compromised, the bearing is well on its way to early failure. A compromised bearing seal encourages dirt ingestion and overlubrication due to its lack of "back pressure" (this is especially true when greasing a bearing "blind" from a remote zirk fitting). The secondary negative effects produced are extra consumption of grease, and extra time required for cleanups of equipment with the overflowed grease, inviting dirt and contaminants to stick to it. *Respect the power of a lowly manual grease gun.*

Maintenance Tip: Always ensure that the dispensing nozzle of the grease or oil gun is cleaned before use, and that the fitting it is being attached to is also clean. This will safeguard against unnecessary introduction of dirt into the bearing.

3.14 How to Change Brands or Types of Grease

The following steps typify the procedure that a maintainer would go through to change out a product brand or grease type.

1. Read and adhere to all old and new grease manufacturer's changeout recommendations.

Case Study VII

THE LETHAL WEAPON!

Understanding the importance of continuous learning is essential for building a sturdy knowledge base. Over the past 15 years, I have delivered many addresses, seminars, and training sessions to hundreds of individuals and companies on industrial lubrication. In order to ascertain the audience's knowledge level early on in the presentation, I ask the audience to tell me how much pressure they think a grease gun delivers. The answers range from 50 psi up to 3000 psi in 95% of all cases! Knowing the simple fact that a good grease gun delivers up to 15,000 psi pressure instantly transforms the audience. Suddenly, the reasons for many different failures are clearly understood: it is at this point that the audience realizes the true importance of skilled lubrication personnel.

2. Isolate equipment (utilize lockout procedures as required).

3. Disconnect lube lines to bearing.

4. Purge each individual bearing of old grease with a hand gun (*not* an air drum pump) containing new grease (open purge plug before commencing). Stop when new grease shows at purge plug or vent. If grease purges past the bearing seals, relube once or twice with small amounts for a few days until new grease appears.

Caution: When using a hand gun, exercise extreme caution in filling, so as not to blow the seal—fill slowly and with a small amount.

5. Purge lube line from opposite end of bearing connection until new grease shows.

6. Disconnect lines at any lubrication blocks, and purge.

7. Purge all old grease from automatic pumping system (if applicable) and fill with new grease.

8. Reconnect all lines and monitor for lube delivery to all lube points.

9. "Soupy" or "hard" grease is evidence of old and new greases still in effect. Repurge as required.

10. Remove lockout.

11. Resume with new recommended lube intervals.

When performing changeout procedures, the supplier of the new grease lubricant should be consulted as to any special procedures that need to be adhered to during the changeout process.

Review Questions

1. Name the three categories of oil classification.

2. Name two advantages and disadvantages of synthetic lubricants.

3. What is viscosity?

4. What is viscosity dependent upon?

5. An oil rated at 1000 SUS@40°C is also known as what rating in the ISO system?

6. What is ISO an abbreviation for?

7. Name two oil additives and their properties.

8. What is grease consistency?

9. Name three types of grease thickeners.

10. What is slumpability?

11. How much pressure can a hand-operated grease gun generate?

CHAPTER 4

How Much and How Often?

4.1 Calculating Bearing Requirements

Over the years, lubricant delivery system manufacturers have devised a basic calculation for bearing surface lubrication requirements. The actual requirements could vary in accordance with the physical conditions found. These outside variables are termed service factors (see Fig. 4.1a).

When lubricating a bearing, the object is *not* to fill the void areas of the bearing, but rather replenish and sustain a lubricant film. This film thickness replacement rate varies, depending on whether the lubricant is oil or grease, and whether it is manually applied or automatically delivered (see Fig. 4.1b). If an automatic terminating grease system were employed, under normal conditions, a 0.001″ (0.025 mm) grease film thickness, applied every 4 hours to the total bearing surface, is all that is required to provide adequate lubrication.

FACTORS AFFECTING LUBRICANT VOLUME REQUIREMENTS

- Machine Condition
- Shock Loading
- Heat
- Speed
- Water
- Environmental Contamination
- Lubricant Type (Oil or Grease)

Fig. 4.1a. Service factors.

Lubricant and Delivery Method	Film Thickness	Time
1. Manually applied terminating grease	0.002" film 0.050 mm film	4 hrs.
2. Automatically applied terminating oil	0.001" film 0.025 mm film	1 hr.
3. Automatically applied terminating grease	0.001" film 0.025 mm film	4 hrs.
4. Automatically applied recirculative oil	0.001" film 0.025 mm film	1 min.

Fig. 4.1b. Replenishment rate of lubrication film under normal conditions.

4.2 Basic Lubrication Requirement Calculation

The basic lubrication requirement calculation, used by the majority of lubrication equipment manufacturers, is

$$V = A \times R$$

where

V = Volume requirement

A = Area of bearing surface

R = Replenishment rate of lubricant film.

Area of Bearing Surface Area: Each type of bearing or bearing surface will require its own calculation. Four types of bearing surfaces are in common use. Calculations are shown in Fig. 4.2a.

Example 1 (all calculations are in inches): A 3" (75 mm)-diameter antifriction, single row ball bearing, running under normal load conditions, is lubricated automatically with a terminating grease system. What would the calculated requirement be in this case?

$$V = A \times R$$

where

$$A = D^2R$$
$$= 3 \text{ in}^2 \times 1$$
$$= 9 \text{ in}^2$$

Plain Bearing: Sleeve, bushing, etc.

$$A = \pi DL$$

where

$\pi = 3.14$
D = Shaft diameter
L = Length of bearing

Antifriction Bearing: Ball, roller, needle bearing, etc.

$$A = D^2 R$$

where

D = Shaft diameter
R = Number of rows

Flat Bearing Surface: Gibs, ways, slides

$$A = \text{Area of contact surface}$$

GEAR (each single gear): Spur, bevel, etc.

$$A = \pi DW$$

where

$\pi = 3.14$
D = Pitch diameter of gear
W = Gear face width

Fig. 4.2a. Bearing surface area calculations.

where

$$R = 0.001 \text{ every 4 hours.}$$

Therefore,

$$V = 9 \times 0.001 = 0.009 \text{ cu. inches/ 4 hours}$$
$$(0.009 \times 16.35 = 0.15 \text{ c.c. (cu. cm) /4 hours.}$$

We can see from this calculation that the amount is indeed very small. The reason for terming these types of bearings "antifriction" is because of their very small bearing surface area.

Example 2: A lubricated bearing block slides over a guide. This block is oil lubricated by a recirculative-type lubrication system. How much oil is required to lubricate under normal conditions? (See Fig. 4.2b.)

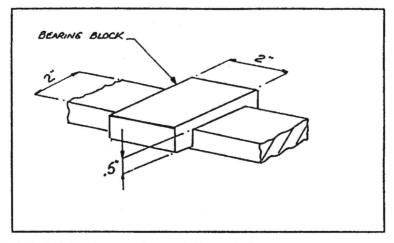

Fig. 4.2b. Lubrication requirement, Example 2.

$$V = A \times R$$

where

$$A = 2 \times 2 = 4$$
$$\text{plus } 0.5 \times 2 = 1$$
$$\text{plus } 0.5 \times 2 = 1$$
$$\text{Total area} = 6 \text{ inches square}$$
$$R = 0.001'' \text{ film every 1 minute}$$

therefore, $V = 6 \times 0.001$

$$= 0.006 \text{ cu. inches every 1 min. or } 0.36 \text{ in}^3/\text{hour}$$
$$\text{continuously or } (0.006 \times 16.35 = 0.098 \text{ c.c.}/\text{min}).$$

When a machine with many lubrication points is considered for an automatic delivery system, the first order of business is to calculate each bearing surface requirement as shown in the examples. The total requirements are added together in order to size a lubricant pump. (The pump is usually sized so that the total amount required falls somewhere in the pump's mid-range capacity. This enables us to cut back or turn up delivery, if needed, without changing pumps, thus ensuring maximum flexibility.) Metering valve sizing and tube sizing are performed according to the system type and the manufacturer's recommendation.

4.3 Exceptions

As previously indicated, these calculations are only guides. Exceptions do occur, and this is where the service factors come into play.

Example: If we place an automatic system on a machine that has already been in use for 5 years, there is bound to be inherent wear on all the bearing surfaces. Lubricant use would therefore be higher than if the machine were brand new.

If wear was a factor, we would calculate the requirements in the standard way shown, and compensation for wear could be adjusted by pump output delivery, pump cycle control (e.g., lubricate every 30 minutes instead of every hour), or both. Because of the many variables involved, it is best to consult the delivery system manufacturer or a consultant for help on these matters.

Another type of exception would be a packaging roller style conveyor table, where the conveyor rollers are held in pillow blocks. Generally speaking, because these bearing loads are so light and intermittent, lubrication other than the original sealed amount is almost never needed.

Each case should be treated on its own merit. Remember, when a machine or bearing manufacturer says it is "lubricated for life," ask *"how long is life?"*

The Danger of Overlubrication: The old saying "a little is good, a lot is better" couldn't be further from the truth when dealing with lubrication—too much lubricant can be as detrimental as not enough lubricant. Chapter 3 explained how viscous shear can cause fluid friction through the wrong choice of lubricant viscosity. Fluid friction is also caused by excess lubricant at the bearing; this is termed "churning," whereby the moving element has to overcome the excess lubricant by pushing it out of the way, thereby causing frictional heat and demanding more energy for the moving element to perform the same work.

Analogy: A simple analogy for overlubrication would be a jog along the beach. If a person were to jog along the beach in one inch of water, the jog would be a very refreshing form of exercise: the kicked-up water would act like a splash lubricant. If, however, this person were to go into twelve-inch-deep water and try to continue the jog at the same pace, the person would very quickly overheat, expending much more energy to overcome the resistance placed by the additional amount of water.

A practical example of overlubrication would be an overhead trolley wheel chain conveyor. On manually greased trolley wheels, the tendency is to completely fill the bearing cavity, thereby creating a potential fluid friction problem. If the amperage draw of the drive motor were monitored when starting up after greasing, we would see a substantial increase in energy consumption, as more power is required to perform the same amount of work (this would also be true if a lubricant were too heavy or viscous). As the lubricant heats up through friction, it will eventually run off or out, and power consumption will return to normal as the bearing cools down.

As we can see, not only is this practice of overgreasing wasteful in terms of grease usage, but it is also costly in terms of power consumption. The effective bearing life is also retarded due to the high temperature "spikes" sustained by the bearing.

Review Questions

1. Name three service factors that will affect lubricant volume requirements.

2. What is the basic lubrication requirement formula?

3. Name four different lubricant and delivery methods that affect the lubricant replenishment rate.

4. What is a common term for fluid friction?

5. How many cubic centimeters make up one cubic inch?

6. When sizing a lubrication delivery pump, what is an important consideration?

CHAPTER 5

Selecting a Lubrication Delivery System

5.1 Oil or Grease?

Whether we have a new equipment design, or a major equipment overhaul requiring a new lubrication policy, the first question to ask is "Should the system use oil or grease?" The decision to use an oil or a grease will be based on the following considerations.

· What type of environment does the equipment reside in? Is the lubricant required to lubricate as well as keep abrasive dirt out from the bearings?

· What temperatures will the bearings be subjected to?

· What is the distance that the lubricant will be pumped? (Pressure drops in grease system lines are much higher than for oil.)

· Will the application utilize manual or automatic delivery, total loss or recirculative methods? (Grease cannot be recirculative—recirculative oil is the preferred method, but it is also the most expensive.)

· Is product contamination a consideration? (Oil is harder to contain than grease.)

· How much has been budgeted for lubrication? (Unfortunately, this is all too often the common denominator and determining decision factor when looking at the design or re-design of a lubrication system.)

5.2 Manual versus Automatic

The essence of good lubrication is to lubricate often, in small amounts. Bearing lubrication requires the *right* lubricant, in the

---

The purpose of automatic lubrication systems:

provide the **right** *amount of lubricant*

at the **right** *location*

at the **right** *time*

by delivering a small amount of lubricant on a continual basis.

Fig. 5.2a. Purpose of automatic lubrication systems.

right place, in the *right* amount, at the *right* time. (See Fig. 5.2a.) Through careful calculation and condition assessment, good lubrication is achievable.

In the majority of circumstances, it is much more beneficial to lubricate with an automatic system. Labor costs are no longer as low as they used to be, and an untrained person armed with a 15,000 psi (1030 bar) grease gun can indeed be a menace! As equipment becomes more complex or increasingly more expensive, it makes sense to spend a very small percentage of the total cost on a very inexpensive insurance policy—a good lubrication system. Not only will it pay back in reduced bearing failures, but it will insure against very expensive downtimes and effectively reduce operating costs. Fig. 5.2b shows the advantages of automatic systems over manual applications. As the figure shows, utilizing an automatic centralized lubrication system allows the bearing to receive a small amount of lubrication on a regular basis, thereby ensuring optimum full film lubrication on a continual basis.

Automatic systems (or centralized systems, as they are also known) have been with us since the early 1920's, when a single line resistance system was first utilized on luxury automobiles such as Packard, Dusenberg, Rolls-Royce, etc. These cars had very demanding lubrication schedules, and the centralized lube system

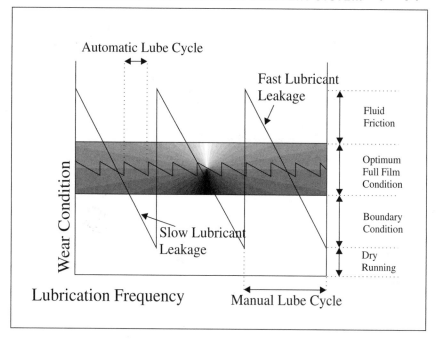

Fig. 5.2b. A reason for choice—manual versus automatic.

served to alleviate that schedule by lubricating all the chassis points automatically, with oil from a single source.

5.3 Delivery Systems

There are many types of lubricant delivery systems available to the purchaser. These range from a grease gun/zirk fitting arrangement, all the way up to a PLC-controlled fully automatic/monitoring delivery system. Most delivery systems are capable of delivering oil in the 100–8000 SSU viscosity range (ISO 22–ISO 1600) and grease up to the NLGI #2 rating. Each method brings its own unique set of characteristics. The following pages will serve as a guide when deciding which system is suitable for the application.

SYSTEM 1:HAND GUN AND ZIRK FITTING

The hand gun and zirk fitting is the simplest and most common method in use. It is also the most dangerous in terms of the damage it can inflict if not used correctly. With grease gun pressures over 15,000 psi (1030 bar) and bearing seal pressures in the 100's psi

Fig. 5.3a. Hand gun and zirk fitting.

(10's bar), the potential for blown seals, overlubrication, and extended lubrication intervals is far too great to use this method successfully. The oiler must be highly trained and regimented. Complex equipment is not tolerant of this method.

ADVANTAGES	DISADVANTAGES
Oil or grease	No metering present
Inexpensive initially	Tendency to overlubricate
No engineering involved	Poor frequency control
	Can damage bearing seals
	Susceptible to dirt inclusion at lube point
	Susceptible to mixing of lubricants

SYSTEM 2: MANUAL CENTRALIZED MANIFOLD SYSTEM

The manual centralized manifold system utilizes the hand gun and zirk fitting method, but this method is advanced due to "tying" all the lube points to a central point via a manifold. Although a time-saving factor is gained, the bearing points are now being lubricated "blind," and can actually compound the problem of overlubrication. Blown seal problems can also be found using this type of application. This method is used on all types of equipment.

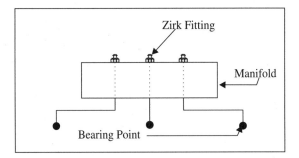

Fig. 5.3b. Centralized manifold system.

ADVANTAGES	DISADVANTAGES
Oil or grease	No metering present
Inexpensive initially	Tendency to overlubricate
No engineering involved	Poor frequency control
Centralization	Can damage bearing seals
	Susceptible to dirt inclusion at lube point
	Susceptible to mixing of lubricants
	"Blind" lubrication

SYSTEM 3: MANUAL PROGRESSIVE LUBRICATION SYSTEM

A hand gun is also utilized with the manual progressive lubrication system, but now a number of lubrication points are "tied" to-

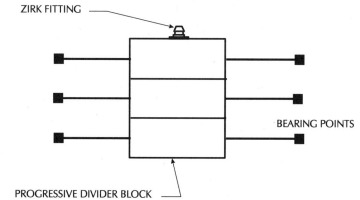

Fig. 5.3c. Manual progressive lubrication system.

gether to a single zirk fitting via a progressive divider block (see Fig. 5.3c for a cross section of progressive blocks). The system is now engineered and eliminates blown seal problems and overlubrication problems with the use of a built-in metering pin which indicates when the lube cycle is complete. This method can be used on all types of equipment.

ADVANTAGES	DISADVANTAGES
Oil or grease	System engineering
Relatively inexpensive	is demanding
Provides a central signal	
Progressive divider block displaces predetermined volumes	
Can be converted to semi- or full-automatic easily	
Positive lubrication	

SYSTEM 4: SINGLE POINT LUBRICATOR

The single point lubricator (see Fig. 5.3d) is a device that screws into the lubrication point. It is a self-contained unit that comes complete with its own reservoir of lubricant (the majority of these reservoirs are refillable).

Lubricant is delivered in different ways. Oil can be metered continually onto a wick or brush which is in contact with the bearing

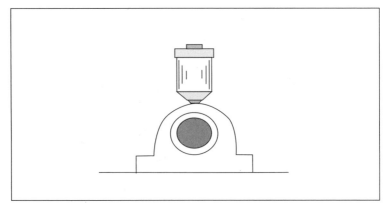

Fig. 5.3d. Single point lubricator.

surface. Grease is applied by a follower plate "pushing" the grease into the bearing. The follower can be spring-loaded, or rely on a chemical or electrochemical chamber which produces a gas that expands, pushing the follower down the reservoir, purging the grease at a variable controlled rate. The expandable gas lubricators are one-time use only and are not refillable.

ADVANTAGES	DISADVANTAGES
Oil or grease	Not always reliable
Inexpensive	Reuseable grease units
Single point application	rely on bearing back
No engineering involved	pressure
No tubing involved	Expandable gas units are
	not always certified for
	use in all areas

SYSTEM 5: SINGLE LINE RESISTANCE SYSTEM (SLR)

The single line resistance system (Fig. 5.3e) is also known as the metering orifice system. This is the forerunner of centralized systems—and the most copied. It can be a full- or semi-(hand-pump) automatic-type system. It is a relatively simple single line system that operates with a low pressure pump [under 200 psi (13.7 bar)], dispensing lubricant via a proportioning or metering orifice. As with

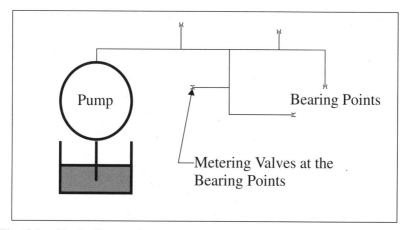

Fig. 5.3e. Single-line resistance.

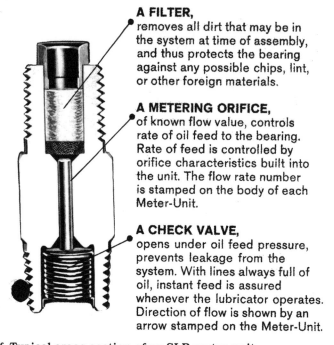

A FILTER,
removes all dirt that may be in the system at time of assembly, and thus protects the bearing against any possible chips, lint, or other foreign materials.

A METERING ORIFICE,
of known flow value, controls rate of oil feed to the bearing. Rate of feed is controlled by orifice characteristics built into the unit. The flow rate number is stamped on the body of each Meter-Unit.

A CHECK VALVE,
opens under oil feed pressure, prevents leakage from the system. With lines always full of oil, instant feed is assured whenever the lubricator operates. Direction of flow is shown by an arrow stamped on the Meter-Unit.

Fig. 5.3f. Typical cross section of an SLR meter unit.

(Courtesy Bijur Lubricating Corp.)

all single line systems, a broken line will neutralize the system (all fluids will take the path of least resistance, i.e., the line break!). The lines are usually small in diameter—5/32″, or 4 mm—and are damaged easily. The orifices are small and can be prone to plugging. (See Fig. 5.3f for a cross section.) Typical applications include vintage automobiles, small machine tools, printing presses, packaging machines, small punch presses, and textile machinery.

ADVANTAGES	DISADVANTAGES
Low pressure applications	Oil only
Easy to add lube points	Not positive lubrication
Generally inexpensive	Does not provide control
Orifices can be built into equipment	signal
System engineering is relatively simple	Small orifices prone to plugging

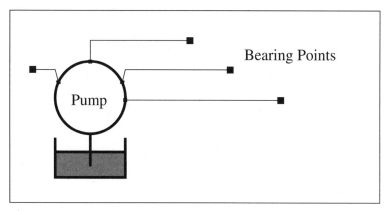

Fig. 5.3g. Pump-to-point system.

SYSTEM 6: PUMP-TO-POINT SYSTEM (MULTIPOINT)

The pump-to-point system (see Fig. 5.3g) can be oil or grease, depending on the pump style and capability. The earlier oil systems were labelled "cam-box lubricators," and were usually shaft driven. The shaft was a camshaft which, when rotated, lifted a rocker arm attached to a small piston device. This in turn pumped oil to the lube point. These piston throws were adjustable, giving some meterability. This type of system is virtually extinct due to the manufacturing costs. Later model pump-to-point systems are much less expensive and contain an air or hydraulic operated pump. The piston chamber contains outlets along its periphery. As the piston actuates and passes by the outlets, lubricant is dispensed to the point. As the piston returns, it draws lubricant back into the chamber, ready for the next shot. The pump-to-point system is restricted to the number of points on the piston wall.

ADVANTAGES	DISADVANTAGES
Oil or grease	Does not provide a central signal
Individual point adjustment	Individual point adjustment
Positive displacement	Tamperable
Easy to engineer	

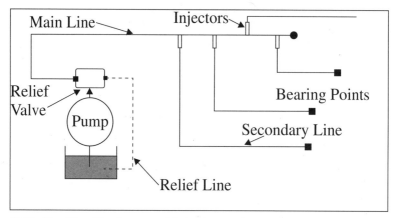

Fig. 5.3h. Positive displacement injector system.

SYSTEM 7: POSITIVE DISPLACEMENT INJECTOR SYSTEM (PDI)

The positive displacement injector system (see Fig. 5.3h) is also known as a single line parallel system. It can deliver oil or grease. It is a single line system that utilizes a pump to force lubricant into the line and injectors, to a pressure of approximately 800 psi (55 bar). (Each lube point requires its own injector—which is usually adjustable by varying the piston stroke.) This pressure can be controlled by an end-of-line pressure switch, or by predetermining and

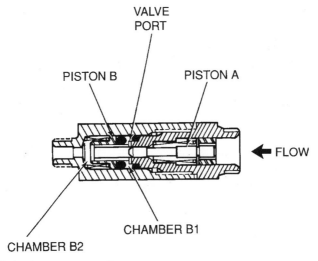

Fig. 5.3i. Injector cross section.

counting the number of pump strokes required to reach required pressure. Individual injectors contain a spring-loaded piston acting against two chambers: a "firing" chamber and a "loading" chamber (see Fig. 5.3i). Once end-of-line pressure is attained, the injector pistons have forced the lubricant in the "firing" chambers to be injected into the lubrication point, while at the same time allowing lubricant into the loading chamber. The pump is now turned off and the line is allowed to relieve itself back to residual pressure by closing the relief valve. As this happens, all injector pistons are spring returned, and lubricant in the "loading chambers" is now pushed into the "firing chamber," ready for the next cycle. This is considered an easy system to design, but is difficult to monitor and can suffer from mechanical failures.

ADVANTAGES	DISADVANTAGES
Oil or grease	Does not provide a central signal
Individual point adjustment	Requires a vent valve and control system
Easy to add lube points	Tamperable
Easy to engineer	

SYSTEM 8: SERIES PROGRESSIVE SYSTEM

Probably the best known type of automatic centralized lubrication system is the series progressive system (see Fig. 5.3j). A pump pumps lubricant—either continually or cyclically—to a network of series progressive divider blocks. These divider blocks are made up of a series of spool valves containing positive displacement pistons. The valves are inter- and cross-ported, linked together in a progressive pattern. The pistons are moved hydraulically by either oil or grease. By attaching a cycle indicator pin to a piston, we are able to easily determine completion of the full block cycle; this is excellent for central control purposes. It is a very flexible system that lends itself well to manual, total loss, and recirculative systems. It is more difficult to engineer in comparison to the single line systems, and adding extra points later is not easily accommodated; but it is a tamper-proof system.

The diagrams, on pages 81–86 (courtesy of Lubriquip Inc.), de-

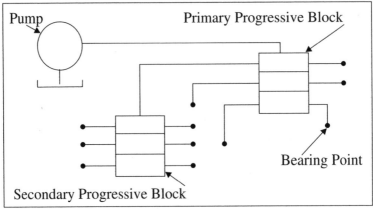

Fig. 5.3j. Series progressive system.

ADVANTAGES	DISADVANTAGES
Oil or grease	Additional lube points
Positive lubrication	require reengineering
Tamper proof	More demanding system
Provides a central signal	engineering

pict a cross section of a standard series progressive divider block, and show the progressive sequence of a standard six-outlet block.

SYSTEM 9: DUAL LINE SYSTEM

The dual line system (see Fig. 5.3k) is also known as the twin line parallel system. It works similarly to the single line parallel sys-

(Continued on page 87)

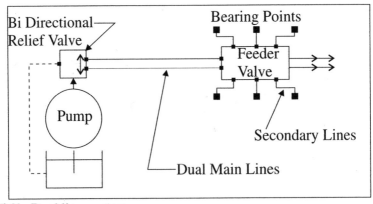

Fig. 5.3k. Dual line system.

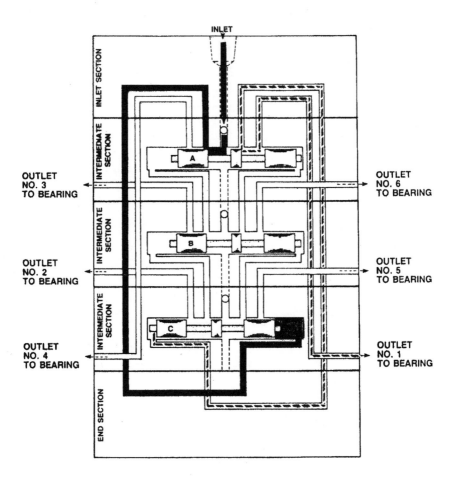

Position No. 1. Lubricant pressure to inlet moves piston C to left forcing a measured amount of lubricant to outlet No.1 bearing.

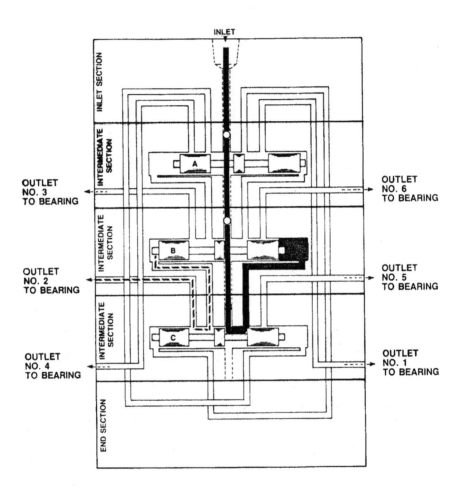

Position No. 2. Movement of piston occurs from C position moving an open channel to the right side of B piston, thus a measured amount of lubricant is forced to outlet No. 2 bearing.

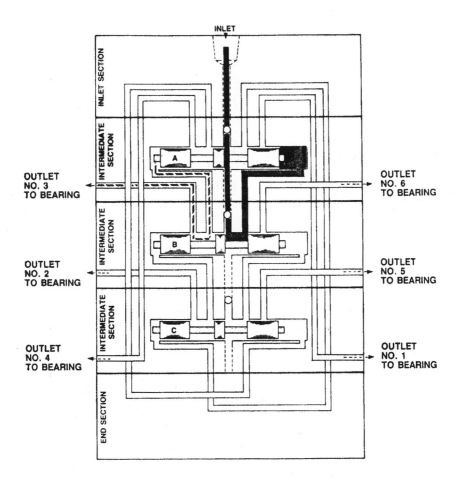

Position No. 3. Movement of piston occurs from the B piston moving an open channel to the right side of A piston, thus a measured amount of lubricant is forced to outlet No. 3 bearing.

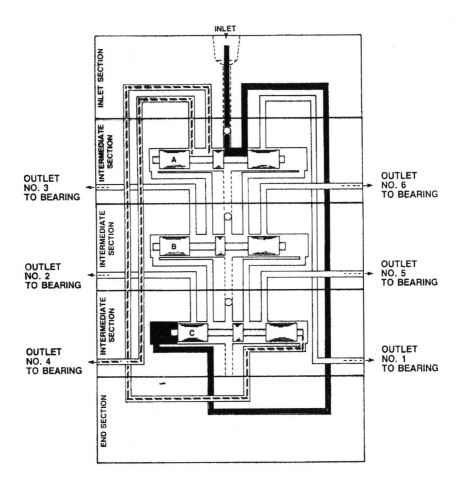

Position No. 4. Movement of piston occurs from the A piston moving an open channel to the left side of C piston, thus a measured amount of lubricant is forced to outlet bearing No.4.

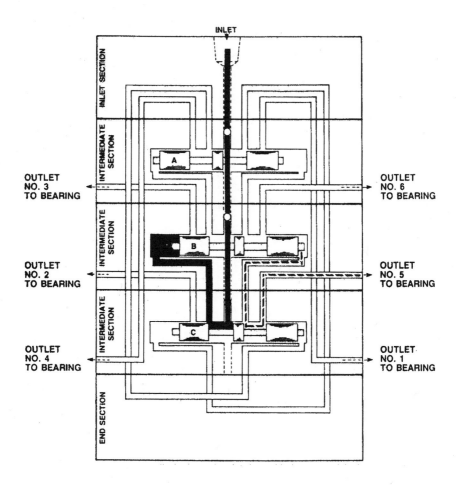

INLET

INLET SECTION

INTERMEDIATE SECTION

A

OUTLET NO. 3 TO BEARING

OUTLET NO. 6 TO BEARING

INTERMEDIATE SECTION

B

OUTLET NO. 2 TO BEARING

OUTLET NO. 5 TO BEARING

INTERMEDIATE SECTION

C

OUTLET NO. 4 TO BEARING

OUTLET NO. 1 TO BEARING

END SECTION

Position No. 5. Movement of the piston occurs from the C piston moving an open channel to the left side of B piston, thus a measured amount of grease is forced to outlet bearing No. 5.

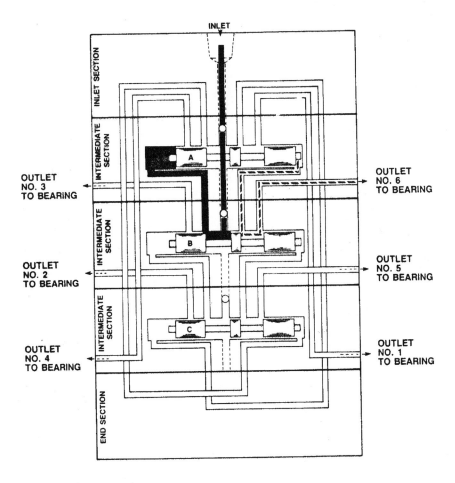

Position No. 6. Movement of the piston occurs from the B piston moving an open channel to the left side of a piston, thus a measured amount of grease is forced to outlet bearing No. 6.

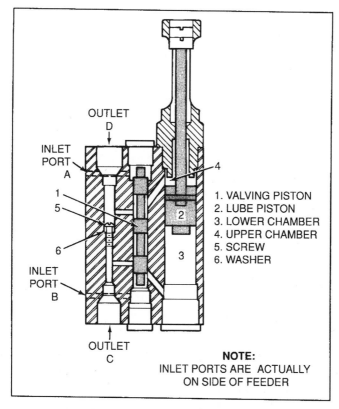

OUTLET
D

INLET
PORT
A

1
5
6

INLET
PORT
B

OUTLET
C

2

3

4

1. VALVING PISTON
2. LUBE PISTON
3. LOWER CHAMBER
4. UPPER CHAMBER
5. SCREW
6. WASHER

NOTE:
INLET PORTS ARE ACTUALLY
ON SIDE OF FEEDER

During feeder operation, pressurized lubricant from the system reverser enters the feeder at port (A) and forces the valving piston (1) down, allowing pressure to be applied to the top of the lube piston (2). Moving down under pressure, the lube piston forces lubricant out of its chamber (3), through the valving piston and outlet (C) to the lubrication point. During this half-cycle of the feeder, the upper chamber of the lube piston was primed for the next feeder operation.

When all feeders have completed the half-cycle, pressure builds at the reverser causing the reverser to shift and direct pressurized lubricant to feeder port (B). The valving piston is forced up allowing pressure to be applied to the bottom of the lube piston. This piston moves up and forces lubricant out of its chamber (4). The lubricant travels past the upper port outlet (D), to the lubrication point.

Cross section of a Trabon® Bi-Flo® dual line lubricant injector.

(Courtesy Lubrigrip Inc.)

tem, with the injectors slightly modified to feed from either side. Line pressure is built up on one side of the system, the pistons fire, the line is then relieved, utilizing a reverser valve, the pump is directed to build up pressure on the second line system, and the sequence repeats itself. For every pumping sequence, half of the system bearings connected are lubricated. This system accepts oil or grease; it is usually the most expensive kind of system, but it can cover extremely large areas with many hundreds of points. Metering is adjustable, therefore tampering is a possibility; but it is a system that is easy to engineer and add points to later.

ADVANTAGES	DISADVANTAGES
Oil or grease	Predetermined valve displacement
Can cover large areas and many points	Adjustable
Easy to add lube points	Doesn't provide control signal
Predetermined valve displacement	Tamperable
Easy to engineer	Expensive installation
Adjustable	

SYSTEM 10: MIST SYSTEMS

Mist systems (see Fig. 5.31) are restricted to oil only. Compressed air is blown into and through a vessel containing lubricant. Through a syphoning process, the air is misted and carried along a header to the lube point. It can then be reclassified to fluid by a special valve at the lube point, or allowed to mist onto the bearing

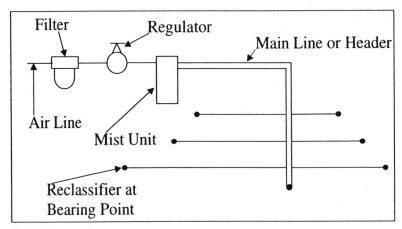

Fig. 5.31. Mist system.

point. Because air and fine droplet formation is involved, good cooling characteristics prevail. However, if the mist is not contained properly, it can cause residual fog drift. In these environment- and health-conscious days, mist lubrication is still used, but only in more controlled situations.

ADVANTAGES	DISADVANTAGES
Continual lubrication	Oil only
Easy to add lube points	Broken line neutralizes
Easy to engineer	system
Provides additional	Difficult to control spray
cooling	mist
	Engineering of distribution system required

SYSTEM 11: AIR/OIL SYSTEMS

An air/oil system is an oil-only system that relies on compressed air (see Fig. 5.3m). Mist lubrication (which also uses air and oil) produces its mist at the reservoir and allows the mist to drift to the lube point. In an air/oil system, the oil is positively metered at the lubrication point and introduced into a constant air stream where it is immediately atomized and forced into the bearing at very high velocity. This is a high-pressure-type system, whereas mist lubrica-

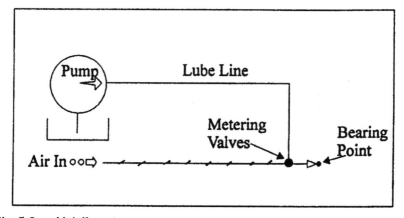

Fig. 5.3m. Air/oil systems.

tion is a low pressure system. Use of air/oil has allowed design engineers to successfully lubricate spindles at very high rpm speeds that were not previously possible.

ADVANTAGES	DISADVANTAGES
Provides additional cooling	Oil only
	Limited application
Allows very high rpm speeds	Requires built-in engineering up front

LUBE LINES

The best material for a lubrication delivery line is steel (in some instances, plastic is a viable alternative). Steel is a little more difficult to work with when installing the lube lines, but it rarely requires replacement, unlike copper tubing. Copper is very easy to install, but is easily crimp damaged, and suffers from work hardening and line splitting. Copper can also deplete certain additive packages from oil. Flexible hoses should be the hydraulic types utilizing compression fittings.

PUMPS

All the lubrication systems discussed in this chapter are different styles of delivery and metering systems. They all rely on having a pump to push the lubricant through the system. Pumps are usually designed to operate in mid-range of adjustability.

Pumps can be hand operated, air operated, hydraulically operated, electrically operated, or mechanically operated. They range in output, size, style, etc., but all manufacturers generally have a pump to suit the application requirements. The most popular style of pump is the piston pump. Other types of lubricant pumps used are gear pumps, vane pumps, and diaphragm pumps.

CONTROLS

System controls are as varied as the pumps. Many third-party control systems are utilized. They can range from a single, stand

Features	Single line resistance (SLR)	Single line parallel positive displacement injector (P.D.I.)	Progressive divider	Twine line parallel dual line system	Pump to point (multi-line)
Oil	✔	✔	✔	✔	✔
Grease		✔	✔	✔	✔
Continuous delivery	✔		✔		
Cyclic delivery	✔	✔	✔	✔	✔
Metering adjustability		✔		✔	✔
Ease of adding on points	✔	✔		✔	
Main line monitor protection	✔	✔	✔	✔	
Secondary line monitor protection			✔	✔	
MACHINERY TYPE N-normal M-medium L-Large	S	S/M	S/M/L	M/L	S/M
ENVIRONMENT N-normal M-medium L-large	N	S.S.	S	S	S
Positive displacement		✔	✔	✔	✔
Approx. number of lubrication points	200 max.	100 max.	200 max.	20 min.	40 max.

This chart is intended as a guide only. Figures may vary from manufacturer to manufacturer.

Fig. 5.3n. Comparison chart for centralized lubrication systems.

alone, clockwork on/off timer, to a sophisticated computer control system built in the machine itself (better known as the PLC or "Programmable Logic Controller"). Choice usually depends on application and budget.

OPTIONS

Many options are available on these systems, depending again on the manufacturer. These include such items as low level alarms, broken line detectors, system failure alarms, overpressure alarms, etc.

Choosing a system will depend greatly on application, number of points, type of control, alarm level required, and budget. Fig. 5.3n shows a comparison chart of the available centralized lubrication systems.

5.4 Basic Selection Procedure for Lubrication System Design

1. Determine if oil or grease is to be used.

2. Select manual or automatic.

3. For automatic, determine pump power source:

   · electric

   · hydraulic

   · mechanical

   · pneumatic.

4. Select control type:

   · visual

   · machine

   · timer

   · count, single, or multiple

   · machine cycle.

5. Calculate bearing requirements.

6 Determine type and pump size.

7. Design meter assemblies (injectors, blocks, units, etc.).

8. Determine system control.

9. Determine options:

- reservoir level indicators

- pressure indicators

- broken line indicators.

Review Questions

1. Name three common factors that would affect the decision to use either oil or grease lubrication.

2. What are the four *rights* of good lubrication?

3. What is the viscosity range of lubricants deliverable by an automatic lubrication delivery system?

4. Name and describe four types of lubricant delivery systems.

5. What are the four methods of powering a lubrication pump?

6. Why is copper tubing not recommended for lube lines?

7. When determining the system requirement, what three kinds of options are looked at?

CHAPTER 6

Preventive/Predictive Maintenance

6.1 Preventive Maintenance

The fact that we are dealing with a preventive maintenance system does not preclude the lubrication system from requiring its own preventive maintenance.

The biggest danger in using automatic lubricating systems is complacency. There is a general tendency to forget to check and fill the lubricant reservoir on a regular basis! This sounds basic, but it is responsible for many failures. Some pumps and injectors are designed to require "priming" upon commissioning. If the lubricant is depleted, the pump loses its prime. If the reservoir is then filled and the pump and injectors are not primed, it appears as if the system is working until someone notices the lubricant level hasn't changed. This could result in serious equipment damage. To overcome this problem, the following can be done:

- ensure that regular fill checks are taking place

- install a low-level switch that will shut down the machine

- clearly signify that the pump is subject to losing its prime

- ensure that lubrication systems checks are a regular part of the preventive maintenance schedule.

It is important to clearly identify the lubricant that is in use. As explained in Chapter 3, mixing lubricants can cause serious problems.

Fig. 6.1 is a preventive maintenance checklist for lubricant delivery systems and can be used for oil or grease systems. When dealing

Description	Grease	Oil
Clean reservoir periodically (do not use cotton or fiber rags).	✔	✔
Inspect suction filter and screens: replace filter at least annually and clean screens.		✔
Remove strainer and clean regularly.	✔	
Change line filter (pressure filter) annually.		✔
Check flexible hoses for cracks and wear.	✔	✔
Inspect tubing/pipe for breaks or flattening.	✔	✔
Check all connections for leaking or "weeping." Check tightness of connection but avoid overtightening.	✔	✔
Monitor system pressure for unusual drops or increases in operating pressure.	✔	✔
Use only recommended lubricants (be cautious about lubricants with additives that could clog filters or flow apportioning devices).	✔	✔
Follow recommended lubricant storage and filling procedures to avoid introducing air and contaminants into the system.	✔	✔

Fig. 6.1. Preventive maintenance checklist for lubrication delivery systems.

with lubrication, we must remember that these systems are not dirt tolerant; therefore, careful attention must be placed on good housekeeping—KEEP IT CLEAN!

6.2 Lube Routes/Color Routes

When setting up for lubrication in a preventive maintenance management system, it is often beneficial to set up the PM task list

Case Study VIII

COLOR ROUTE SYSTEMS ASSIST LUBRICATION PRACTICES

As can be seen in the Alemite color route card, color route management systems have been in use for decades. Modern management philosophies often adopt and update these tried and tested methods. Total Production Maintenance (TPM) philosophy pays much attention to lubrication procedures and their benefits. TPM also relies on operator involvement.

This study involved a chemical process facility that was changing over to a TPM philosophy. In order to train operators to perform effective lubrication without information overload, the color identification system was proposed. All grease guns were to be standardized for output delivery and then colored for the type of grease in use.

All manual lubrication points requirements are then assessed and calculated. The points are then hooked into a manual series progressive block with a single grease nipple inlet. This block is colored according to the type of grease, and a tag plate is attached to the block engraved with the number of grease gun strokes required for the operator to perform correct lubrication.

Operators are comfortable with this method because the subjectivity of the function has disappeared. This method also encourages "ownership" of the equipment by the operator.

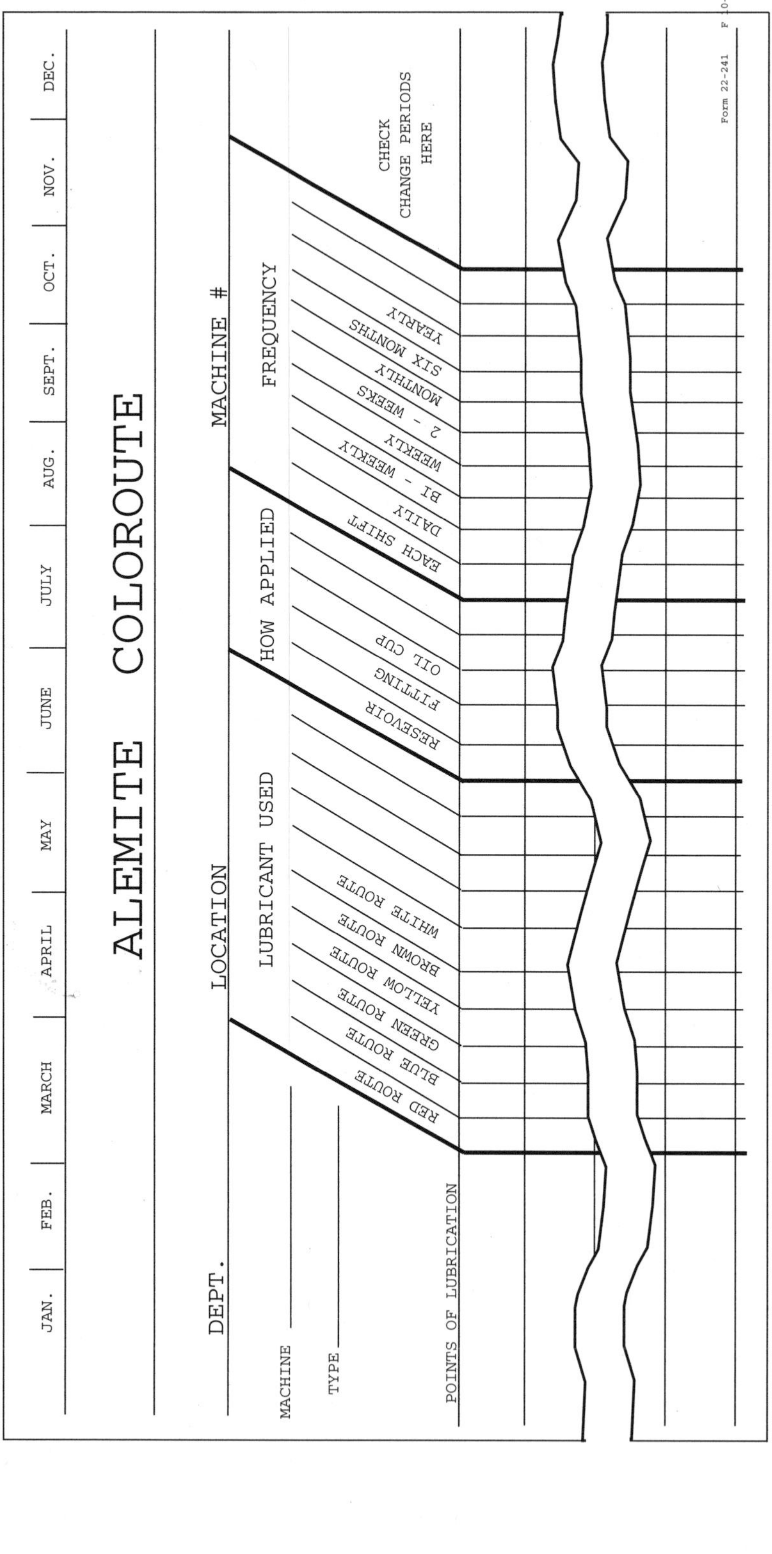

Fig. 6.2. ALEMITE "COLOROUTE" chart (Stewart-Warner Corp.).

to follow a specific route. Routes can be set up as a specific color, e.g., orange route.

Example: The orange route could be set up to use the orange grease gun which contains XYZ grease, where the lubricator will pass by all identified machinery that contains orange-marked grease points and deliver the specified amount of grease to all the points.

This type of system enables machine operators to perform basic lubrication without fear of bearing damage, and adapts well to Total Productive Maintenance (TPM) environments.

6.3 Wear Particle Analysis—Predicting Lubricant Failure

Wear Particle Analysis (WPA) is a predictive maintenance technology used exclusively for lubricants. WPA is the most common method of testing oils and greases for contamination or deterioration. WPA is an inexpensive method of testing: an oil sample is taken from a lube oil or hydraulic oil reservoir and sent to a testing laboratory. WPA (or "oil analysis," as it is often termed) is probably the most cost-effective tool in the technological toolbox. This is largely due to the fact that no major capital outlay is needed in order to get started.

WPA traces its roots back to the 1940's, when the North American railways decided to analyze their electromotive diesel engine oil; in particular, they were looking for premature engine wear by studying the amount of silver bearing wear particulate in ppm (parts per million). The U.S. Air Force quickly followed, utilizing the technique for monitoring their aircraft engine wear. By the 1960's, independent laboratories had enabled private industry to enjoy the many benefits of WPA.

There are basically two detection methods employed in oil analysis—ferrography and emission spectrometry. Both methods involve taking an oil sample from the source to be monitored (this is usually performed with a syringe or suction gun device; care must be taken to only draw lubricant into the suction tube, and once transferred into the sample container, discard the old suction tube and replace it with a new suction tube ready for the next sample). The laboratory performs tests and sends the results back to the client. A good laboratory will also trend the results against previous samples from the same source and flag significant changes.

FERROGRAPHY

Using an instrument known as a ferrograph, the wear particles are separated magnetically from the oil onto an inclined glass slide. The particles are distributed along the slide according to their size. The ferrogram, as it is now known, is then treated to enable the particulate to adhere to the surface once the lubricant is removed. Using an analysis technique known as bichromatic microscopic examination, the particulate density and ratio size are then studied for indication of the extent and type of wear that has taken place. An electron microscope is sometimes used as an alternative instrument for the study.

This method is very accurate and more sensitive than emission spectrometry for detecting the early stages of wear. It actually measures the particulate shape and size.

SPECTROMETRY

Spectrometry is divided into two methods—emission spectrometry and atomic absorption spectrometry. Emission spectrometry excites the metal particulate or elements in the sample by applying a high voltage charge, usually 15,000 V or more. This causes the impurities to emit a characteristic radiation which can be measured and analyzed. In atomic absorption spectrometry, the sample is diluted and vaporized with an acetylene flame. Because atoms absorb light at different wavelengths, we are able to detect and quantify the presence of wear elements by emulating their light wavelengths. Atomic absorption spectrometry is probably the most common method of detection employed today.

OTHERS

In addition to checking the standard wear elements (see Fig. 6.3a), a variety of other physical tests can be performed, dictated by the lubricant's application. A few common ones are as follows.

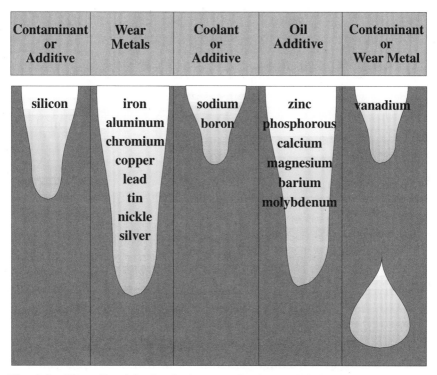

Contaminant or Additive	Wear Metals	Coolant or Additive	Oil Additive	Contaminant or Wear Metal
silicon	iron aluminum chromium copper lead tin nickle silver	sodium boron	zinc phosphorous calcium magnesium barium molybdenum	vanadium

Fig. 6.3a. Wear Particle Analysis—standard tests. This chart shows the standard elements tested for in a wear particle analysis test.

· *Particle Count:* This is performed on hydraulic oils where it is important to know the micron size and count of the particulate in particles/ml. Hydraulic oil manufacturers publish recommended levels of particulate in order to gauge the lubricant's cleanliness.

· *Water, Glycol, Fuel:* Tests are performed to determine the presence of these. Depending on the application, their presence may be desirable or undesirable. For example, if a motor oil were being tested, the presence of water, glycol, or fuel could indicate a faulty head gasket. Alternatively, if a check were being run on a coolant, then water is desirable and its percent content is required.

· *Viscosity:* The lubricant's viscosity is checked in order to identify it as the correct lubricant, and to check if it is still within specification.

Case Study IX

WEAR PARTICLE ANALYSIS (WPA) PAYS OFF

We have shown how WPA plays an important role in determining root cause analysis of failure of the air tools and hydraulic valves. Oil samples that show high concentrations of wear metals, dirt, glycol, water, or abnormal conditions indicate the presence of wear and accelerated failure. The sample in this study was taken from a large drive unit in a brewery.

A good lab will show a "trendline" on each sample. When a lube sample program is in place, sampling of components is performed on a regular basis. Because of this regular sampling, any "trend" that occurs can be closely monitored. The trending was the direct factor that saved the drive unit.

The sampling showed that the copper readings had increased over three samplings from 11 to 133 to 147 ppm, consecutively. This was enough evidence to warrant a full internal inspection of the unit. The unit was found to contain badly worn bearings. Because sampling had taken place on a regular basis, the worn bearings were found just before the drive unit's warranty had expired. This enabled the brewery to claim the $5,000 repair expense under warranty.

Case Study X

WEAR PARTICLE ANALYSIS (WPA) PAYS OFF

The sampling was taken from a rotary screw air compressor. In this particular study, the client was basing oil changes on the WPA results. The analysis results showed dirt levels above normal and an oil change was necessary. The source of dirt was of concern; and airborne chemical products were determined as another source of contamination shown by the level rise in calcium.

This particular client is a manufacturer of detergent powders. In this type of manufacturing process, calcium is part of the manufactured product. Airborne calcium contaminants were present; this was supported by the ppm level rise of calcium.

To solve the problem of dirt and calcium ingestion, the compressor air inlet ducting was moved to a cleaner location and a prefilter was put in place at the inlet mouth. Extended lubricant life and a cleaner running, more efficient compressor was the result. This is typical of the many ways in which WPA is utilized on an everyday basis for predicting failure, and is used as a direct means for finding equitable solutions to everyday plant problems.

 WEAR PARTICLE ANALYSIS

EngTECH Industries Inc

240 Holiday Inn Dr., Suite R.,
Cambridge, Ontario N3C 3X4
Tel/Fax: (519) 661-0311

ISO CODE	PARTICLES/M1.	ISO CODE	PARTICLES/M1.
4	0.16	14	160
5	0.32	15	320
6	0.64	16	640
7	1.3	17	1280
8	2.5	18	2560
9	5	19	5120
10	10	20	10240
11	20	21	20480
12	40	22	40960
13	80	23	81920

ENGTECH INDUSTRIES INC,
240 HOLIDAY INN DR. SUITE 'R'
CAMBRIDGE
ONTARIO   N3C 3X4

FLEET UNIT M-Q PUMP P.P.   FUEL
COMPONENT HYDRAULIC   MADE BY
MODEL   SERIAL#
OIL BRAND PETRO   SAE GRADE   AW 32

NORMAL = N, ABNORMAL = A, SEVERE = S

DIAGNOSIS: UNIT WEAR RATES APPEAR IN THE NORMAL RANGE. PARTICLE COUNT IS UNACCEPTABLE. REFILTER OIL OR CHANGE OIL AND FILTERS TO CLEAN OUT SYSTEM. RESAMPLE AT HALF THE NORMAL INTERVAL TO MONITOR.

P/O # 43980

DATE	05/03/91	15/04/91	21/11/91	01/07/92	01/08/92	13/05/93	01/10/93	Trendline
Mi/Hr UNIT	0	0	0	7230	7805	8583	1259	
Mi/Hr OIL	1920	1920	240	3650	7117	7692	5852	8160
OIL ADD	Y	Y	N	N	N	N	N	
ALUMINIUM	1 N	0 N	1 N	1 N	2 N	1 N	2 N	
CHROMIUM	0 N	0 N	0 N	0 N	0 N	0 N	0 N	
COPPER	14 N	3 N	41 A	67 N	58 N	34 N	54 N	:X
IRON	4 N	2 N	5 N	9 N	9 N	5 N	20 N	
LEAD	0 N	0 N	0 N	0 N	1 N	1 N	2 N	
NICKEL	1 N	0 N	0 N	2 N	0 N	0 N	1 N	
TIN	1 N	0 N	2 N	0 N	0 N	2 N	1 N	
SILICON	1 N	0 N	0 N	0 N	0 N	0 N	2 N	
SODIUM	6 N	4 N	4 N	9 N	6 N	6 N	14 N	
BORON	8	2	14	11	9	3	0	
ZINC	426	418	406	504	471	418	408	
PHOSPHORUS	339	313	351	426	386	360	339	
MAGNESIUM	6	3	3	6	3	4	13	
CALCIUM	98	109	141	172	150	87	100	
BARIUM	17	30	25	32	30	23	12	
MOLYBDENUM	0	0	0	0	0	0	4	
VIS 40 C.	38	36	35	35	34	33	32	
WATER	34	37	48	128	137	96	112	
T.A.N.	.43	.41	.43	.41	.38	.40	.40	
>5u	100000	37500	9120	TOO	2680	62	35809	
>15u	38000	4160	980	CONTAM	290	8	4563	
>25u	3500	106	45	INATED	26	1	1223	
>50u	120	12	8	TO	4	0	142	
>100u	31	3	3	COUNT	1	0	6	
ISO CODE	24/22	22/19	20/17	TCTC	19/15	13/10	22/19	
CODE #	Z151	Z174	Z174	H17	H4	N2	H4	
LAB. #	521	521	183	372	167	193	85	

Fig. 6.3b. Typical spectrographical oil analysis.

(Courtesy Engtech Industries Inc.)

· *Neutrality Number:* This can be a TAN test to check the Total Acid Number or acidity. When used in conjunction with the viscosity rating and particulate presence, we are able to determine if oxidization is occurring.

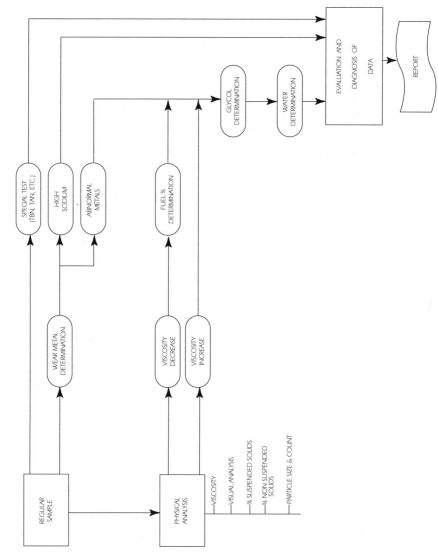

A TYPICAL LABORATORY FLOW CHART FOR A ROUTINE OIL ANALYSIS

Fig. 6.3c. A typical laboratory flow chart for a routine oil analysis.

The obvious benefit derived from wear particle analysis is the ability to predict component failure by checking for and analyzing the component material found in the lubricant. Secondary benefits are derived by checking the lubricant's health. For example, if the manufacturer advises that the lubricant be changed out every 3 months, and a check shows no deterioration or additive drop out, we can extend the oil changeout period to a suitable period dictated by the machine's actual operating conditions, thereby saving on lubricant costs and downtime costs.

The spectrographical oil analysis sheet depicted in Fig. 6.3b shows what the WPA user may receive from the laboratory.

Review Questions

1. Why is complacency so dangerous when utilizing auto lube systems?

2. Lubrication systems are dirt tolerant. True or False?

3. Name three preventive maintenance checks for oil.

4. Name three different preventive maintenance checks for grease.

5. In Wear Particle Analysis, what is a particle count used for?

6. Name a benefit of Wear Particle Analysis.

7. Name five elements tested for in Wear Particle Analysis.

CHAPTER 7

Health, Safety, Storage, and Handling

7.1 Health and Safety

When introducing lubricants into the workplace, it is vitally important to obtain the lubricant manufacturer's Material Safety Data Sheet (MSDS). The MSDS contains the information required to safely handle and store the product. If any information is unclear, the lubricant vendor should be contacted for clarification.

The MSDS sheet is broken into nine (9) sections, as follows.

Section 1—Product Identification and Use
The actual product and its intended uses are identified along with the manufacturer's and supplier's names and addresses.

Section 2—Hazardous Ingredients
Harmful and hazardous ingredients are listed.

Section 3—Physical Data
Information regarding state, odor, appearance, specific gravity, boiling and freezing points, etc., are identified.

Section 4—Fire and Explosion Data
This section indicates the product's flammable characteristics and means of extinction.

Section 5—Reactivity Data
How the product reacts chemically with other substances is noted here.

Section 6—Toxicological Properties
This part tells us if the product is toxic and how it will affect people who are exposed to it.

Section 7—Preventive Measures

These are the measures that dictate the type of protection required for handling and storage of the product.

Section 8—First Aid Measures

This section indicates the measures necessary for first aid treatment.

Section 9—Preparation Date of the MSDS

Different states and provinces will have legislation in place that will mandate the use of MSDS information. It should be kept readily available for any person to see at any time.

It is a good practice to keep a set of MSDS sheets "off-site" in case of fire: the fire department can be notified of the types of toxins they are facing in the fire-fighting process.

7.2 Lubricant Storage Containers

There are many styles of lubricant containers available; and your usage will determine what standard containers your lubricant is shipped in. Certain lubricants require specialized handling and storage procedures—consult your supplier for proper procedures when purchasing new lubricants.

If you have many different lubricants in present use, a lubrication audit could determine if the lubrication requirements could be consolidated in order to substantially reduce stocking requirements. The wide flexibility offered by today's lubricants allows this type of effort to work well.

Fig. 7.2 shows the typical container sizes available for oil and grease. When transferring lubricants, it should be remembered that

Oil	Grease
5 gallon pail	35 lb. Pail
16 gallon drum	120 lb. Pail
45 gallon drum	400 lb. drum
bulk tank-tote	bulk tank-tote

Fig. 7.2. Common lubricant container sizes.

the same transfer pump should not be used for different lubricants. Lubricant properties are easily changed by adding another kind of lubricant to it. This also applies to funnels and small handling containers.

Special Notes:

· Typically, 5- and 16-gallon refinery containers for oil are an open top design.

· Similarly, all grease refinery containers are an open top design.

· 45-gallon oil drums are bung hole design.

· Bulk tanks are available in top–bottom–side mounts.

· There are 42 U.S. gallons in one barrel of oil.

Conversion Volume:

To convert U.S. gallons to Imperial gallons: × 1.2104
To convert U.S. gallons to DM3(liter): × 3.78541
To convert pounds to kilograms: × 0.45359

7.3 Storage Dos and Don'ts

DO

· Keep containers tightly closed to avoid dust and dirt contamination.

· Store containers in a place which will allow containment of leaks to avoid entry into sewers and ground.

· Store containers away from areas of spark or flames in a covered area with adequate ventilation.

· Heed manufacturer's labeling (e.g., some manufacturers require lubricant containers to be stored on their sides in racks).

DON'T

· Never store drums or pails outside without protection from the elements. Temperature fluctuations often cause a low pressure in the container which will allow water to be drawn through even the tightest fitting cap, thus contaminating the lubricant.

· Never store containers in extreme heat or cold.

7.4 Handling and Disposal Dos and Don'ts

DO

· Wear goggles and viton/butyl rubber gloves when pouring and handling lubricants.

· Use a respirator if pouring or handling lubricants in a confined or poorly ventilated area for prolonged periods.

· Store waste lubricants in proper containers, making sure not to mix volatile or hazardous liquids with the waste lubricants, as it will dramatically escalate disposal costs.

· Arrange for disposal of waste lubricants by a reputable, licensed carrier.

· Post MSDS sheet regarding special handling instructions.

· Wash hands before handling food.

· Develop a spill action plan.

DON'T

· Never use dispensing pumps for different products unless thorough cleaning and purging is done before switching products (i.e., drum and pail pumps).

· Never dispose of lubricants directly into sewer or ground.

· Never siphon lubricants by mouth.

Review Questions

1. What is an MSDS sheet?

2. How many U.S. gallons are in one barrel of oil?

3. Name two Dos of lubricant storage.

4. Name two Don'ts of lubricant storage.

5. Name two Dos of lubricant handling and disposal.

6. Name two Don'ts of lubricant handling and disposal.

CHAPTER 8

Filtration

8.1 Filter Classifications

There are two basic types of filter element design employed in lubrication delivery systems—surface and depth (see Fig. 8.1).

SURFACE-TYPE FILTERS

A surface filter separates unwanted particulate and debris from the lubricant through a "screening" process. The filter medium "screens" the unwanted material onto a common single surface. The surface filter medium is typically:

- wire mesh
- nylon mesh
- pleated paper
- membrane
- magnetic
- porous metal.

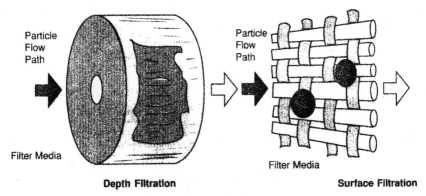

Depth Filtration   **Surface Filtration**

Fig. 8.1. The two basic types of filter design.

(Courtesy STLE.)

In a lubrication system, surface-type filters are used as elementary or "first stage" filtration. They are usually found in the oil reservoir acting as a suction screen filter on the pump inlet line. They are also found on the fill opening of the reservoir where they resemble a mesh sock. A magnetic drain plug is a reservoir that would also be considered a surface-type filter.

Surface filters are also found on the pressure side of the lubricant pump in the form of a spin-on-type paper filter (automatic style) or replaceable paper/porous metal elements. Because the oil only flows through or across a single surface, these types of filters are sometimes referred to as "short flow path" filters. They are used for filtering out coarse particulate from 40 micron size and above.

Example: A good example of a surface filter found in the home is the everyday coffee maker filter.

DEPTH-TYPE FILTERS

The construction of a depth filter is such that the oil must take a torturous path through a much denser filter medium. The lubricant "resides" in the filter medium much longer than with a surface filter, and the filter medium is able to filter out much finer particulate matter. A depth filter medium is typically:

· cotton waste

· wool

· felt or flannel

· shredded paper

· wound paper.

All of the above materials have good absorbent qualities (these materials are also known as "inactive").

Depth filters are more commonly found on hydraulic systems as an optional parallel circuit to the surface filter. 10%–15% of the lubricant is "bypassed" through the depth filter for "deep cleaning." The system effectiveness can range from 40 to 1/2 micron.

Example: A good example of a depth filter material found in the home resembles a roll of toilet tissue.

8.2 Filter Ratings

Filters are rated by manufacturers in microns:

- one micron or micrometer = one millionth of a meter

- one micron or micrometer = 0.00004 inch

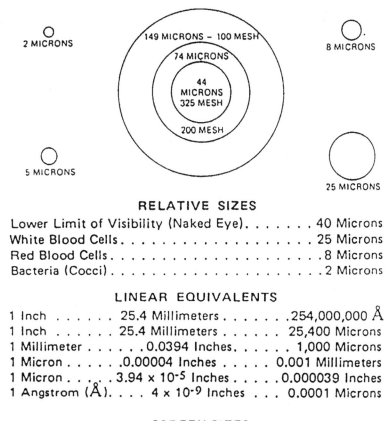

RELATIVE SIZES

Lower Limit of Visibility (Naked Eye). . . . . . . 40 Microns
White Blood Cells. . . . . 25 Microns
Red Blood Cells. . . . . .8 Microns
Bacteria (Cocci). .2 Microns

LINEAR EQUIVALENTS

1 Inch . . . . . . 25.4 Millimeters . . . . . . .254,000,000 Å
1 Inch . . . . . . 25.4 Millimeters . . . . . . 25,400 Microns
1 Millimeter . . . . . . .0.0394 Inches. . . . . . 1,000 Microns
1 Micron . . . . . .0.00004 Inches . . . . . 0.001 Millimeters
1 Micron . . . . . .3.94×10^{-5} Inches . . . . .0.000039 Inches
1 Angstrom (Å). . . . 4×10^{-9} Inches . . . 0.0001 Microns

SCREEN SIZES

Meshes Per Linear Inch	U.S. Sieve No.	Opening in Inches	Opening in Microns
52.36	50	.0117	297
72.45	70	.0083	210
101.01	100	.0059	149
142.86	140	.0041	105
200.00	200	.0029	74
270.26	270	.0021	53
323.00	325	.0017	44
		.00039	10

Fig. 8.2. Micron size relationships.

The naked eye's lower limit of visibility is 40 microns or 0.0016 inch. (See Fig. 8.2 for more information on microns.)

Maintenance Tip: When changing out the lubricant, *always* replace the filter. This will ensure full system design pressure and full life expectancy of the system and components.

Review Questions

1. Name three types of filter material used in "surface" filters.

2. Give an example of a "surface" filter use.

3. Name three types of filter material used in "depth" filters.

4. Give an example of a "depth" filter use.

5. What is a full flow filtration system?

6. What is a micron?

7. What is the smallest micron size visible to the human eye?

Glossary of Terms

additive: Any material added to oil or grease to improve its performance in service. Additives are designed to enhance the lubricant's properties or give it properties it would not otherwise possess.

acid: A chemical substance characterized by having an available reactive hydrogen requiring alkali to neutralize it.

AGMA: American Gear Manufacturers Association. One of its activities is the establishment and promotion of gear lubricant standards.

alkali: A chemical substance having marked basic properties. The term is applied to hydroxides of ammonium, lithium, potassium, sodium, borium, and calcium. It is soluble in water and has the power to neutralize acids.

antifoam agent: An additive which causes foam to dissipate more rapidly. It helps to turn small bubbles into large bubbles which burst more easily.

antifriction bearing: A bearing employing rollers or balls—usually termed rolling bearings.

antioxident agent: A chemical additive which increases a lubricant's oxidation resistance, which lengthens its service and storage life.

antiwear agent: An additive used to minimize wear caused by metal-to-metal contact during conditions of mild boundary lubrication (i.e., stop–starts, oscillation motion, etc.).

API: American Petroleum Institute.

ASME: American Society of Mechanical Engineers—an organization and governing body devoted to furthering excellence in the field of mechanical engineering.

ASTM: American Society for Testing and Materials—an organization devoted to the promotion of knowledge of engineering materials and standardization of specifications and testing methods used in lubricant manufacturing.

base stock or base oil: The refined mineral or synthetic fluid into which additives are blended to produce a finished product.

bleeding: See **oil separation.**

boundary lubrication: A state of lubrication that occurs when, due to a combination of speed, load, and lubricant property, the lubricant film has not developed sufficiently to prevent the metal surfaces from contacting each other. Special additives are sometimes used for bearing protection under these conditions.

cavitation: In a lubricant delivery system, this is a condition occurring when reduced pressure causes a void (or cavity) impeding suction and preventing lubricant flow (usually at the pump inlet).

centipoise/centistoke: A centipoise is 1/100th of the unit of absolute viscosity, the *poise.* The viscosity of water at 20°C is approximately one centipoise. The centipoise is derived from one kinematic unit of viscosity, the centistoke. 1 centipoise = 1 centistoke × density of liquid. These units are metric.

channeling: Formation of a groove in grease or heavy oil. Channels are cut by the motion of a lubricated element (i.e., gear or rolling contact bearing), leaving shoulders of lubricant serving as a seal and reservoir. This action is usually desirable, although a channel too deep or permanent can cause lubrication failure.

chatter: When meshing gears are not lubricated, the mating teeth set up shock waves; this is heard as a "chatter"-like sound.

churning: An action resulting when a moving element has to "plow" its way through an overlubricated cavity. This action causes fluid friction and heats up the bearing.

cloud-point: The temperature at which paraffin wax or other solids begin to crystallize or separate from the solution, giving a cloudy appearance to the oil.

color route: See **lubrication route.**

compatibility: A lubricant's ability to be mixed with another lubricant without detriment to either lubricant.

complex soap: A soap wherein the soap crystal or fiber is formed by co-crystallization of two compounds, i.e., normal soap (such as metallic stereate or oleate), and the complexing agents, which usually increase the grease's dropping point.

consistency: A basic property describing the hardness or softness of grease.

Chemical Wear (CW) lubricant: The most common class of lubricant in use today. To prevent mechanical seizure, these lubricants coat the metal surfaces with aggressive chemicals that "soften the surface." If the metal surfaces do collide, the asperities are allowed to break away easily, causing minimal damage to the host material.

degree Engler: A measure of viscosity. The ratio of time for the flow of 200 ml of test liquid, through the viscometer designed by Engler, to the time required for the flow of 200 ml of water, equals the degree Engler.

demulsibility: A lubricant's ability to separate from water.

depth filter: An extremely "dense" filter where the lubricant is allowed to reside in the filter media for an extended time to allow a filtration level down to 1/2 micron. Often found in hydraulic systems, depth filters are commonly known as "deep cleaning" filters.

detergent: An additive in crankcase oils generally combined with dispersant additives. It chemically neutralizes acidic contaminants in the oil before they become insoluble and drop out of the oil, forming sludge. The dispersant agents operate to break up the insoluble particles already formed. Particles are kept finely divided to remain colloidal within the oil.

DIN: Deutsche Industrie Norm—a German national standards governing body.

dispersant: A dispersing agent, compatible with the carrier fluid, which holds a finely divided third-party substance in a dispersed state in the carrier fluid.

drop point: The temperature at which grease changes from semi-solid to liquid under test conditions; often an indication of temperature limitation for application purposes.

EHD lubrication: Elasto-hydrodynamic lubrication is predominantly found in rolling element bearings. EHD is a hydrodynamic film formed by applied pressure or load. The pressure increases between the rolling element and the raceway of the bearing as the ball or roller comes into the load area; this pressure on the lubricant "squeezes" the lubricant into a thin hydrodynamic lubricant film that is almost "solid" in nature. (See Hertzian.)

emulsifier: An additive used to promote or help emulsification of two liquids and enhance its stability.

EP (Extreme Pressure) agents: Additives added into lubricant to prevent sliding metal surfaces from seizing under extreme pressure conditions. At these high local temperatures, the agents combine chemically with the metal to provide a surface film that prevents welding and scoring. Compounds of sulphur, chlorine, and phosphorus are often added for this purpose.

"F" lubrication factor: The relationship between the degree of lubrication protection and lubricant film thickness.

ferrography: A popular method of oil analysis which detects ferrous, nonferrous, and nonmetallic particles, both large and small, that could be suspended in the lubricant.

fire point: The temperature rating at which a lubricant will catch fire.

flash point: Minimum temperature of a petroleum product at which vapor is produced at a sufficient rate to yield a combustible mixture. It is the lowest sample temperature at which the mixture will "flash" in the presence of a small flame ignition.

friction: The force which opposes the movement of one surface sliding or rolling over another with which it is in contact.

grease: A lubricant composed of a lubricating fluid, thickened with soap or other material, to a semi-solid or solid consistency.

Hertzian: When a load is applied to a rolling element bearing, the point of contact between the rolling element and the raceway will elastically deform. This contact region is called the Hertzian contact area or Hertzian region. (See EHD lubrication.)

hydrodynamic lubrication: This is caused solely by the "pumping" action developed by the sliding of one surface over another in contact with a lubricating oil. Adhesion to the moving surface draws the oil into a high pressure area between the surfaces, and viscosity retards the tendency to squeeze it out. If the pressure developed by this action is sufficient to separate the mating surfaces, full-film lubrication is said to take place.

inhibitor: An additive used to control undesirable chemical reactions in all lubricants; e.g., oxidation inhibitors, rust inhibitors, foam inhibitors, etc.

IP: Institute of Petroleum (U.K.).

ISO: International Standards Organization—a worldwide standards governing body, gradually becoming the preferred standard rating for viscosity.

Jost Report: A landmark study, commissioned by the British Government in 1966. The study committee, headed by Peter Jost, researched and reported on how friction, lubrication, and wear directly and indirectly affected the country's Gross National Product (GNP) in the areas of industry, natural resources, and agriculture.

lubricant: A substance (for example, grease, oil, etc.) that, when introduced between solid surfaces which move over one another, reduces resistance to movement, heat production, and wear by forming a fluid film between the two surfaces.

lube route: A lubrication task list that follows a "mapped"-out route of lubrication points to be lubricated in a single work session. Different routes can be designated by color (see color route).

maintenance: The act of preserving or keeping equipment and facilities in the proper condition.

maintenance ready: Equipment that has been designed to optimize and facilitate the maintenance function.

mineral based lubricants: Mineral based lubricants utilize base stocks refined from natural crude oil.

MSDS: Material Safety Data Sheets. All lubricants are supplied with MSDS information which gives the user vital information regarding the safe handling and storage of the lubricant.

NLGI: National Lubricating Grease Institute. Numbers are given to different consistency grades of grease using the ASTM D-217 cone penetration test.

nonsoap thickener: A substance such as clay, silica gel, carbon black, or any of several specially treated or synthetic materials that can be thermally or mechanically dispersed in liquid lubricants to enable formation of lubricating grease.

oil separation: Oil separation occurs in grease when the oil "separates" or "bleeds" from the grease thickener. This usually occurs when the grease has not been "worked" for a while.

PD lubricant: Plastic Deforming (PD) lubricants are third-generation lubricants. They are usually mineral based lubricants that contain liquid metallic additives. If friction occurs, the ensuing high temperatures activate the metallic additives which then penetrate the working surface and allow that surface to soften and deform without metal loss.

PDI: Positive Displacement Injectors are a type of lubricant metering device used in automatic lubrication delivery systems. The injectors use a positive displacement piston.

PLC: Programmable Logic Control.

PM: A common acronym that can mean: Preventive Maintenance, Predictive Maintenance, Proactive Maintenance, Planned Maintenance, Productive Maintenance, or Profit Maintenance.

pour point: A widely used low temperature flow indicator which is 5°F (−15°C) above the temperature at which a normally liquid product maintains fluidity.

pour point depressant: A lubricant additive that allows the lubricant to remain fluid in extremely cold conditions.

pumpability: A grease's ability to flow easily under pressure through a distribution system over a given temperature range.

rust: A residue (usually reddish brown) that forms on the surface of metal, which is caused by oxidation.

rust inhibitor: A lubricant additive for protecting ferrous components from rusting caused by water contamination or other harmful materials from oil degradation (sludges).

ROI: Return On Investment.

SAE: Society of Automotive Engineers.

saponification: This is the process where fat or fatty acids react with an alkali to form a soap. This process is the basis of grease manufacturing.

shear stability: The ability of a grease to resist changes in consistency (hardness) during mechanical working. Working may be in one of several laboratory machines, or in actual service.

shearing: The action that allows friction to occur between the molecular planes of the lubricant film instead of the metal surfaces.

SLR: Single Line Resistance is a type of automatic lubrication delivery system that relies on proportional metering valves to dispense the lubricant to the lubrication point.

slumpability: A property of some greases that makes them partially self-levelling within containers. A desirable condition when using central lubrication systems.

soap: A specific type of salt formed by the reaction of a fatty acid with an alkali.

solid lubricant: A class of lubricant where friction reduction is caused by making the shearing take place within the crystal structure of a material with low shear strength in one particular plane; e.g., graphite, molybdenum disulfide, and certain soaps. Grease is not a solid lubricant, but may contain solids as additives.

spectrometry: The most common method of oil analysis. Spectrometry checks for up to 18 different elements in the lubricant. The differing levels of these elements will indicate wear and what type of wear is occurring with the lubricated component, as well as the level of contamination of the lubricant.

SSU: Saybolt Second Universal viscosity measure. The time (in seconds) for 60 ml of fluid to flow through a Saybolt Universal Viscosimeter at 104°F and 210°F.

STLE: Society of Tribologists and Lubrication Engineers—an institution dedicated to promoting and furthering the science of tribology/lubrication.

surface filter: A screen-type filter that "screens" the unwanted material onto a common single surface.

SUS: Saybolt Universal Second (see SSU).

synthetic lubricants: These lubricants possess a base oil that has been manufactured from chemical constituents or by polymerization of hydrocarbon (olefins) rather than by conventional refining of petroleum. The three most common types of synthetic base oil are: polyalphaolefins, organic esters, and polyglycols.

TAN: Total Acid Number is a measure of the increased acid level in the lubricant. It usually refers to industrial lubricants such as hydraulic and gear oils.

TBN: Total Base Number is a measure of the reserve alkalinity remaining in engine lubricants.

TPM: Total Productive Maintenance is a maintenance philosophy that focuses on good lubrication, good housekeeping, and operator involvement in the maintenance process.

tribology: The science of lubrication, friction, and wear.

viscosity: The measure of a fluid's resistance to flow. Ordinarily expressed in terms of the time required for a standard quantity of lubricant at a certain temperature to flow through a standard orifice. The higher the value, the more viscous is the fluid.

viscosity improver: A polymeric additive used to increase the viscosity index (see V.I.) of an oil; it is used extensively in multigrade-type oils.

Viscosity Index (V.I.): The measure of the rate of change of viscosity with temperature. This change is common to all fluids, but does not occur at the same rate. The higher the V.I., the less the tendency for viscosity to change.

weeping: The action of a lubricant bleeding through a porous casting or a lubricant line that has work hardened at a bend point and has become porous.

worked penetration: The penetration level of a grease sample immediately after it has been subjected to 60 strokes in a standard grease worker.

WPA: Wear Particle Analysis is a predictive maintenance technology that analyzes the particulate matter in oil and grease. It is often termed "oil analysis" (see ferrography and spectrometry).

zirk fitting: The correct term for a "grease nipple." The fitting is usually found at the lubrication point. It is shaped to accept a grease gun nozzle and allow lubricant to flow through it one way. The internal ball check valve also serves to keep dirt out when not in use.

ASTM Routine Tests for Lubricants

ASTM—American Society for Testing and Materials

Unworked and Worked Penetration (ASTM D217)—NLGI Grade: Numerical value of lubricating grease consistency (soft greases have high penetration values and low NLGI classification grades).

Thickener Type: Instrumentation, such as atomic emission and dispersion infrared spectroscopy, identify grease base which relates to application functions.

Specific Gravity, Weight per Gallon (ASTM D1298, D1963): Ratio of density of lubricant to density of water weighed in air. Significantly high density or lbs/gallon value can reflect more product consumption by weight if volumetrically metered in lube system.

Dropping Point (ASTM D2265): Temperature at which grease passes from semi-solid to liquid and, therefore, no longer functions as thickened lubricant. Normally, high drop point greases can be used at proportionately higher temperatures.

Viscosity—Kinematic (ASTM D445, D2161), Apparent (ASTM D2983): Measure of resistance to flow of lubricant at a given temperature (higher value typically denotes more thin film strength).

Viscosity Index (ASTM D2270): Empirical number indicating effect of temperature change on oil viscosity. High V.I. signifies a relatively small change in viscosity.

Flash/Fire Point (ASTM D92, D93): Temperature at which lubricant vapors initially ignite (flash) and subsequently burn (fire). Generally, a higher temperature corresponds to lower volatility.

Nature of Applied Film w/o Solvent: Thin film structural tenacity can be proportionate to lubricant's protective ability.

Water Washout (ASTM D1264); Water Sprayoff (ASTM D4049): Direct measure of lubricant resistance to water attach measured in percent loss.

Rust (ASTM D1743, modified D665): Indicates corrosion-rust preventative properties of lubricant. Lower number equals better result; no corrosion/rust has #1 rating/pass.

Lincoln Ventmeter, psi @ degrees F: Shows lubricant's ability to flow in lube line at a given temperature. Lower psi result denotes both less ventmeter viscosity and flow resistance, pointing to better pumpability.

Solvent-Carcinogenic Determination: Instrumentation, such as Fourier Transform Infrared spectroscopy (FTIR), identifies solvents. U.S. and international agencies, such as the National Toxicology Program (NTP) and International Agency for Research on Cancer (IARC), list known/suspected carcinogens (cancer causing), and evaluate carcinogenic risk of chemical to humans.

Roll Stability, Pts. Change (ASTM D1831): Measure of grease's ability to resist change in consistency during mechanical working/shear (indicative of directional change (soften–harden) in consistency of grease during service).

Prolonged Worked Penetration, 10M-100M Strokes (ASTM D217): Shows resistance of grease to shearing action. Higher numerical difference between worked and prolonged worked penetrations denotes less shear stability.

Wear; Four Ball (modified ASTM D2266, D4172), Falex (ASTM D2670), Falex Model I Continuous Load Simulation: Shows wear-preventing properties of lubricant. Small Four Ball scar, low Falex teeth number, and high pound-force equates to better antiwear protection.

Extreme Pressure: Four Ball (ASTM D2596, D2783), Falex (ASTM D3233, D2625); Timken (ASTM D2509, D2782), Falex Model I FZG Simulation: The greater the Four Ball Weld and Index values, the better are the extreme pressure, antiwear properties. Higher Falex pound-force and/or load stages and Timken pound-numbers equate to better load carrying qualities.

Laboratory Examination: Arc emission spectroscopy and high-powered optical microscopy show numerous elements, including both additives and contaminants such as lead, zinc, calcium, iron, molybdenum, copper, sodium, etc.

Evaporation Loss (Modified ASTM D972, D2595): Evaluates volatility characteristics of lubrication based on time/temperature. This is an important factor in maintaining lubrication and efficiency of unit/components.

Conradson Carbon Residue (ASTM D189): Determination of amount of carbonaceous residue formed after evaporation/pyrolysis of oil. High value correlates to more deposit/coke forming tendencies of lubricant.

Oxidation Stability (ASTM D942, D943): Ascertains resistance of lubricant to oxidation. Lower value denotes less formation of volatile oxidation byproducts and better stability under static conditions.

Foam (Modified ASTM D892): Rates oil's resistance to foaming. A high tendency to foam (fail rating) can lead to mechanical failure of a unit from inadequate lubrication, cavitation, and system overflow.

Demulsability (ASTM D1401): Establishes ability of oil to separate from water after emulsification. Normally, a short time for separation represents ultimately better protection by lubricant under water-contamination conditions.

Copper Corrosion (ASTM D130, D4048): Assesses degree of corrosivity of lubricant to copper. Lower number means better result; slight corrosiveness/tarnish has number 1A rating.

Lubricating Solids: A combination of lubricating solids, physically and chemically bonded, effects a synergistic thin film barrier between metal surfaces; this results in reduced friction, increased load carrying capabilities, and antiwear protection.

Appendix of Commonly Used Conversion Factors

To Convert From	To	Multiply By
atmospheres	inches of mercury (32°F)	29.921
atmospheres	pounds/sq ft	2116.32
atmospheres	pounds/sq inch	14.696
barrels, oil	gallons (U.S.)	42
BTU (60°F/15.56°C)	joule	1055
centimeters	feet	0.0328
centimeters	inches	0.3937
cubic centimeters	cubic inches	0.0610
cubic centimeters	gallons (U.S.)	0.00026
cubic centimeters	ounces (British, fluid)	0.0351
cubic centimeters	ounces (U.S., fluid)	0.0338
cu ft of water (60° F)	pounds	62.37
cubic inches	cubic cm	16.3872
feet	meters	0.3048
gallons (British)	cubic cm	4546.08
gallons (British)	gallons (U.S.)	1.2009
gallons (British)	liters	4.5459
gallons (U.S.)	cubic cm	3785.434
gallons (U.S.)	liters	3.7853
grams	ounces (avoirdupois)	0.03527
grams/liter	parts per million (ppm)	1000
horsepower	foot-pounds/second	550
horsepower	watts	745.7
inches	centimeters	2.54
inches of mercury (32°F)	atmospheres	0.0334
inches of mercury (32°F)	feet of water (39.2°F)	1.133
kilograms	ounces (avoirdupois)	35.274
kilograms	pounds (avoirdupois)	2.2046
kg-meters (torque)	pounds-feet	7.2330
kilometers	feet	3280

To Convert From	To	Multiply By
kilometers	miles	0.6213
kilowatt-hours	BTU	3413
liters	cubic inches	61.025
liters	gallons (British)	0.2199
liters	gallons (U.S.)	0.2641
meters	feet	3.2808
meters	inches	39.37
miles	feet	5280
miles	kilometers	1.6093
milliliters	cu inches	0.061
milliliters	ounces (British, fluid)	0.035
milliliters	ounces (U.S., fluid)	0.0338
millimeters	inches	0.039
ounces (avoirdupois)	grams	28.3495
ounces (British, fluid)	cu cm	28.4130
ounces (U.S. fluid)	cu cm	29.5737
pounds (avoirdupois)	grams	453.5924
pounds/sq ft	kg/sq meter	4.8824
tons (metric)	kilograms	1000
tons (metric)	pounds (avoirdupois)	2204.62
tons (short)	kilograms	907.1848
tons (short)	pounds	2000

Bibliography

ASME Research Committee on Lubrication (1981), Strategy for Energy Conservation Through Tribology, 2nd ed., American Society of Mechanical Engineers, New York.

Bannister, K. E. (1989), "Lubrication and Maintenance Practices," Engtech Industries Inc., Maintenance Management Seminar Series.

Bannister, K. E. (1990), "Automation Answers for Lubrication Problems," *Plant Engineering and Maintenance Magazine,* A Clifford Elliot Publication, Canada

Bannister, K. E. (1992), "Wear Particle Analysis is Effective and Economical," *Plant Engineering and Maintenance Magazine,* A Clifford Elliott Publication, Canada.

Canada National Research Council (1986), "A Strategy for Tribology in Canada—Enhancing Reliability and Efficiency Through the Reduction of Wear & Friction," NRC Ottawa.

Interlube International, Inc. (1987), "Lubrication Technical Manual—Blue Book," Interlube International Inc., WA, licensed under Optimol Olwoerke, Munich.

Jost, H. Peter (1966), "Lubrication (Tribology)—A Report on the Present Position and Industry's Needs," U.K. Dept. of Education & Science, HMSO, London.

Shell International Petroleum Co. Ltd. (1964), "Lubrication of Industrial Gears," Shell International, London.

Stewart, Harry L. (1981), "Pumps," Audel, Indiana.

SKF Canada (1990), "SKF Ball & Roller Bearing Refresher Booklet."

Society of Tribologists & Lubricating Engineers (STLE) Alberta Section (1990), "Basic Handbook of Lubrication," p. 24.

Texaco/Shell/Petro-Canada/Esso/ICI-Tribol/Optimol Olwoerke, Products Digests.

Trabon/Bijur/Lincoln/Farvel, Products Digests.

Index

Index

A

Acid, 33, 48, 53, 104, 117
Additive(s), 24, 25, 48, 49, 62, 101, 117
AGMA (American Gear Manufacturing Association), 40, 42, 117
Air/oil, 89
Aluminum, 53, 54, 58
Animal/vegetable oil, 31
Antiwear, 30, 48, 117
Antiweld, 34
Aromatic hydrocarbon, 33
Asperities, 10, 13
ASTM (American Society for Testing and Materials), 47, 52, 53, 56, 121, 125
Automatic lubricating systems, 39, 41, 66, 67, 70

B

Barium, 53, 54, 58
Base (oil) stock, 33, 34, 44, 48, 118
Bearing(s), 14, 15, 18, 19, 22, 23, 39, 40, 51, 63, 65, 68, 70, 75, 81, 82, 83, 84, 85, 86, 92, 102
Bentone, 54, 58
Bleeding, 55, 118
Boron Nitride, 24
Boundary, 19, 22, 43, 118

C

Calcium,3, 54, 58, 103
Cavitation, 118
Centralized Lubrication System, 54, 55, 70, 91
Centralized manifold, 72, 73
Changeout, 26, 29, 34, 38, 48, 51, 58, 59, 61
Channelling, 118

D

E

F